高等院校土建学科双语教材（中英文对照）
◆ 建筑环境与能源应用工程专业 ◆

BASICS

# 室内环境调节
# ROOM CONDITIONING

[德] 奥利弗·克莱因　　著
　　 约尔格·施伦格尔
马志先　　　　　　　译
张吉礼　　　　　　　审

中国建筑工业出版社

著作权合同登记图字：01-2009-7701号

**图书在版编目（CIP）数据**

室内环境调节/（德）克莱因，（德）施伦格尔著；马志先译．—北京：中国建筑工业出版社，2014.4
高等院校土建学科双语教材（中英文对照）建筑环境与能源应用工程专业
ISBN 978-7-112-16390-8

Ⅰ．①室… Ⅱ．①克…②施…③马… Ⅲ．①室内环境—环境设计—双语教学—高等学校—教材—汉、英 Ⅳ．①TU238

中国版本图书馆CIP数据核字（2014）第027063号

Basics: Room Conditioning / Oliver Klein, Jörg Schlenger
Copyright © 2008 Birkhäuser Verlag AG, P. O. Box 133, 4010 Basel, Switzerland
Chinese Translation Copyright © 2014 China Architecture & Building Press
All rights reserved.
本书经Birkhäuser Verlag AG出版社授权我社翻译出版

责任编辑：孙书妍　　责任设计：陈　旭　　责任校对：陈晶晶　刘梦然

高等院校土建学科双语教材（中英文对照）
◆建筑环境与能源应用工程专业◆
**室内环境调节**
[德] 奥利弗·克莱因　著
　　　约尔格·施伦格尔
马志先　　　　　　译
张吉礼　　　　　　审
＊
中国建筑工业出版社出版、发行（北京西郊百万庄）
各地新华书店、建筑书店经销
北京嘉泰利德公司制版
环球印刷（北京）有限公司印刷
＊
开本：880×1230毫米　1/32　印张：$5\frac{1}{4}$　字数：170千字
2014年10月第一版　2014年10月第一次印刷
定价：18.00元
ISBN 978-7-112-16390-8
（25140）

**版权所有　翻印必究**
如有印装质量问题，可寄本社退换
（邮政编码100037）

## 中文部分目录

\\ 序　_5

\\ 导言　_88

\\ 设计原理　_90
    1.1　舒适性需求　_90
    1.2　确定需求　_97
    1.3　补进需求　_104

\\ 通风系统　_118
    2.1　自然通风　_119
    2.2　机械通风　_125
    2.3　选择正确的系统　_130

\\ 调温系统　_132
    3.1　能源供应　_133
    3.2　热量与冷量存储　_143
    3.3　热量与冷量分配　_143
    3.4　热量与冷量传递　_145
    3.5　选择正确的系统　_154

\\ 通风与调温的结合　_156
    4.1　可行方案的范围　_156
    4.2　选择标准　_156

\\ 结语　_157

\\ 附录　_158
    \\ 一些概念的案例　_158
    \\ 标准　_166
    \\ 参考文献　_167
    \\ 作者简介　_167
    \\ 译者简介　_167

# CONTENTS

\\Foreword _7

\\Introduction _8

\\Design principles _10
    \\Comfort requirements _10
    \\Determining the requirements _17
    \\Covering the demands _24

\\Ventilation systems _38
    \\Natural ventilation _39
    \\Mechanical ventilation _45
    \\Deciding on the right system _50

\\Tempering systems _52
    \\Energy supply _53
    \\Heat and cold storage _63
    \\Heat and cold distribution _63
    \\Heat and cold transfer _65
    \\Choosing the right system _74

\\Combination of ventilation and tempering _76
    \\The range of possible solutions _76
    \\Selection criteria _76

\\In conclusion _77

\\Appendix _78
    \\Examples of concepts _78
    \\Standards _86
    \\Literature _87
    \\The authors _87

# 序

确保人们的安全和健康是建筑非常重要的功能。建筑保护人们远离天气的波动及其带来的不利影响。同时，必须给室内提供足够的清新空气和热量或冷量以保证舒适度。因此，室内空间环境的调节是建筑的一个重要方面，并且它远远超出了纯粹的供暖与通风系统的技术实施。明智的设计及它们的实现可以将建筑结构、功能与技术全面地联系在一起，以实现降低或完全消除建筑能量需求这一目标。

为能够从建筑设计一开始就考虑这一指导性原则，需要对利用室内环境调节来满足需求的各种要求与可能性有丰富的认识。这包含多种技术体系，尤其是对它们之间的相互作用与相互依存关系的理解。因此，将拟定室内环境调节的构思视为初步设计的组成部分是非常重要的。

《室内环境调节》使用易于理解的介绍和解释，一步一步详细地阐述了这一主题。这本书适合处在其职业生涯开端的学生和专业人员使用。首先，这本书介绍了多种基本的舒适性需求，并基于功能、应用和气候条件等阐述了它们之间的区别。用于确定需求与满足需求的原则表明，为避免引起不利的环境效应并保持最小的能量需求，必须在项目规划的早期阶段设立后续技术实现的方向。

其他章节讨论了机械通风和自然通风概念，并为系统的选择提供了应用背景。从能量的供应和可能的存储，到能量在建筑中的分配和能量在室内的传递，以及最终调温系统与通风的关系，系统地阐述了室内温度调节。

作者提供了对室内环境调节的相互关系和可能性的充分、广泛的认识。为了使用这些知识实现每栋建筑的最佳的室内环境调节概念，最重要的考虑因素不是技术体系的数理设计，而是明智与综合的方法。

编辑：Bert Bielefeld

# FOREWORD

Ensuring people's protection and wellbeing are important functions of buildings. They preserve people from fluctuations in the weather and from its adverse effects. At the same time the interior must be provided with sufficient fresh air and heat or cold to guarantee comfort. The conditioning of interior space is therefore an important aspect of architecture, which extends far beyond the purely technical implementation of heating and ventilation systems. Intelligent designs and their realization link building structure, function and technology holistically, with the aim of reducing a building's energy demand or eliminating it entirely.

A wide knowledge of the requirements and possibilities of covering demand with room conditioning is needed to be able to take this guiding principle into account from the outset when designing a building. This includes the technical systems and, above all, an understanding of the interactions and interdependencies. It is important to see drawing up a room conditioning concept as an integral part of preliminary design.

*Room Conditioning* elaborates this subject step by step using easily understood introductions and explanations. It is intended for students and professionals at the start of their careers. First, the book presents the fundamental comfort requirements and differentiates between them according to function, use and climatic conditions. The principles used in determining and covering demand make clear that the direction for the subsequent technical implementation must be set in the early stages of project planning, to avoid adverse environmental effects and to keep energy demand to a minimum.

Further chapters discuss mechanical and natural ventilation concepts, and provide a practical context for system finding. Room tempering is explained systematically, from energy supply and possible storage, through to its distribution in the building and transfer in the room, and finally its relationship with ventilation.

The authors provide a sound, broadly based understanding of the interrelationships and possibilities of room conditioning. The most important consideration is not the mathematical design of technical systems, but an intelligent and integrated approach in order to use this knowledge to achieve an optimum concept for every individual building.

Bert Bielefeld, Editor

# INTRODUCTION

**The temperature of the human body**

While many species are able to adjust their body temperatures to suit their surroundings, humans require an almost constant body temperature of 37 ± 0.8 °C. As the outside temperature fluctuates depending on the climate zone, time of the day and season, the human body attempts to maintain this temperature using an automatic regulation system, in which the surface of the skin gives off more or less heat, according to the ambient temperature and the level of physical activity. For example, if the body temperature rises, sweat glands allow moisture to emerge onto the skin, where it evaporates and gives off heat into the environment. If the body temperature drops, the skin contracts to reduce the area giving off heat, and the hairs on the skin stand up ("goose pimples"). The body also creates additional heat by quivering its muscles (shivering).

**Climatic influences and compensation**

However, this temperature regulation system also has its limits, and the human skin can only fulfill this task to a certain extent. Clothing as additional "thermal insulation," the "second skin," and buildings, the "third skin," provide the solution.

In human history, the discovery of fire was surely the most important step for humankind in achieving independence from climatic conditions and the seasons. It was not only the entry into the fossil fuel age, i.e. energy conversion dependent on a continuous supply of energy sources; the "third skin" could then have artificial heat and light – and thus it was also the original form of room conditioning. Today's problems of environmental destruction associated with the use of fossil fuels are well known to us and omnipresent.

**Energy-optimized room conditioning**

The term "room conditioning" is understood as the creation of an indoor climate for people to enjoy a feeling of wellbeing, which is above all unaffected by any outside influences by tempering (heating or cooling), lighting and the introduction of sufficient fresh air (ventilation); with suitable technology, this can lead to an ultimately uniform architecture that is unrelated to its location. In extreme cases, these buildings are hermetically sealed with glass facades, fully air-conditioned by the extensive use of high technology, and can be found built to practically the same design in all climatic regions of the world. In addition to disturbing user sensitivities, another disadvantage is the very high energy requirement for heating, cooling and lighting. The fact that 50% of the total energy consumed worldwide is used in buildings shows that other ways must be devised to provide energy-optimized room conditioning.

A building should always be designed to provide comfort using only a small amount of additional energy. First, all available constructional (passive) measures for room conditioning should be exploited, taking into account local conditions, before turning to technical (active) measures.

› Chapter Design principles

The effective combination of passive and active measures in which all the technical components are mutually compatible is always crucial to obtaining an energy-optimized overall concept for room conditioning. The following chapters explain the basic principles and their various interactions, and are intended to enable an individual and balanced room conditioning system to be developed for every building project.

# DESIGN PRINCIPLES

## COMFORT REQUIREMENTS
### Thermal comfort

The term "comfort" describes a feeling of wellbeing, which is influenced by a number of factors. In the field of building services systems, it generally refers to thermal comfort, describing a state in which the body's thermal balance is in equilibrium with the climatic conditions of the surroundings. The user perceives the climatic conditions of the surroundings as neither too hot nor too cold.

<small>Importance of comfort</small>

Thermal comfort is not a luxury; it is an important criterion for being able to use a building fully for its intended purpose. The quality of the space in a building has many different effects on the ability of its occupants to concentrate and work effectively, as well as their state of health (e.g. in offices). If comfort is inadequate in production areas, this may give rise to premature fatigue with the corresponding consequences for safety at work. The reliable provision of an indoor climate appropriate to use is therefore an important quality characteristic of a successful building concept.

### Influence factors

Our perception of comfort depends on a number of influence factors, which are shown in Figure 1.

The designers of a building are normally only able to influence physical conditions, some of which are described in detail in the following chapters. However, the clothing and activity of the user clearly affect his or her perception of comfort. Along with the user's ability to adapt and acclimatize, both belong to a group of "intermediate" factors and are influenced by both physical and physiological conditions.

In some cases, other factors may also play an important role in the preliminary building design, the knowledge of which and the conscious creation of a building concept specifically for a defined user group are often indispensable for the design process. For example, older people often find higher temperatures comfortable, which must be taken into account by providing appropriately higher room air temperatures when designing a nursing home for the elderly.

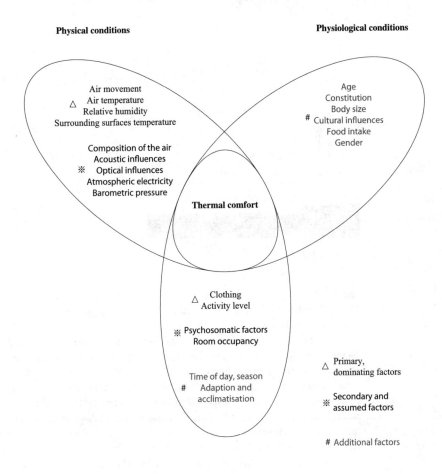

Fig.1:
Factors influencing thermal comfort

## Physical conditions

*Air and radiation temperature*

After air temperature, the most important physical factor is the average temperature of the surrounding surfaces. Like any other body, the human body is also continuously exchanging heat with surrounding surfaces by means of radiation. Depending on the distance and the

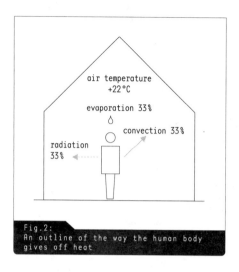

Fig. 2:
An outline of the way the human body gives off heat

> 🏠

> ✏️

Operative temperature

temperature difference between two bodies, more or less heat will be given off or taken up in one or the other direction. This process affects the body's heat balance.

Slight differences between air and radiation temperatures may still be perceived as comfortable by the human body. If the difference between air and radiation temperatures, or even the difference between radiation temperatures, is too great, this causes discomfort. This is why you feel uncomfortable standing near a very warm or very cold radiating surface (e.g. a poorly thermally insulated building component or window), despite pleasant room air temperatures. > Fig. 3

As the human body cannot detect absolute temperatures, only more or less intense heat loss or gain by the skin, people's temperature

\\ Note:
With little physical activity, normal clothing and ordinary room temperatures, one third of the heat the human body loses is through radiation, convection and evaporation (see Fig. 2).

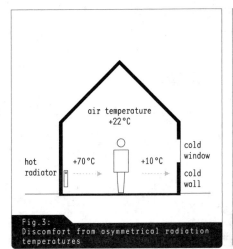

Fig.3:
Discomfort from asymmetrical radiation temperatures

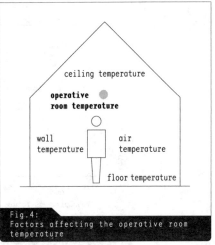

Fig.4:
Factors affecting the operative room temperature

sensitivity depends on the exchange of heat with the air and the surrounding surfaces in the room. The combined effect of air and radiation temperatures on the human senses is expressed by the "operative temperature" (or "perceived temperature"), which has become accepted as an authoritative design parameter for assessing comfort.

The operative room temperature can differ in various parts of the room depending on the distance to the room surfaces. For design purposes, it is calculated as the average air temperature and the average radiation temperature of all room surfaces, and is used to calculate the thermal state of the room.

Humidity

A further factor affecting our perception of comfort is the relative humidity of the air, because the human body gives off part of the heat

\\ Tip:
With predominantly sedentary activities, e.g. in offices or homes, ceilings that are too warm or wall or window surfaces that are too cold quickly cause discomfort. The temperature difference between surfaces and room air should not be more than 3 Kelvin (K). Within certain limits, higher air temperatures can compensate for low surface temperatures and vice versa.

\\ Note:
The recommended ranges of operative temperatures vary depending on use and also in international comparisons. For Europe, according to European standard EN 15251, the recommended range for light, sedentary activities is about 20-26 °C. Many other counties have their own regulations, which often differ from this range.

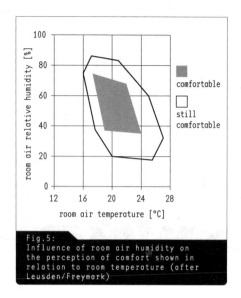

Fig. 5:
Influence of room air humidity on the perception of comfort shown in relation to room temperature (after Leusden/Freymark)

through evaporation. Depending on the humidity, a given temperature state with otherwise identical boundary conditions may be perceived as warm or cold. › Fig. 5

In addition to moisture loads (people and plants in the room), another influence on the relative humidity in the room is the outdoor air humidity (i.e. the climate). Relative humidity also varies with fluctuations in room air temperature. Thus, warming of cold outdoor air in winter normally results in lower relative humidity indoors.

\\ Note:
A relative humidity of about 50% (± 15%) is often recommended for most uses (e.g. light sedentary activities in homes or offices). However, it must taken into account that very few room conditioning systems (e.g. full air conditioning systems) allow the user to have complete control over relative humidity.

**Air movement**

Air movement can have a considerable influence on thermal comfort independently of the humidity. Draughts can be caused by air moving at excessive speeds and by eddies (high degree of turbulence in the air flow). The perception of comfort depends on the temperature of the air: the lower the air temperature, the sooner a movement of air is perceived as uncomfortable, while a movement of warm air is tolerated for rather longer. The higher the air temperature, the less strong eddies are a problem.

**Composition of the air**

Although the composition of the air at ground level is generally constant, it is nevertheless able to influence the overall perception of comfort. This is because of the widely varying levels of pollution and temperature-dependent moisture content of the air that can occur in particular situations.

Air pollutants can arise from outside and inside a building. While the outdoor influences are mainly a function of the building location (road traffic etc.) and the positions and sizes of the ventilation openings, the principal indoor influences can be identified as the emissions of the building (vapors given off from the materials used, other materials in the building or uses of the building), as well as the people present in the building (carbon dioxide content of their exhaled air, water vapor, odors etc.). Eliminating these emissions and/or providing adequate air exchange should enable any specified limits for the various substances in the air to be observed and discomfort – or damage to health – to be avoided.

**Carbon dioxide content of the air**

The carbon dioxide content of the room air is particularly significant. It is added to by the breath of the people in the room and only has to reach a few parts per 1000 before it becomes a problem. Contrary to popular opinion, it is not a lack of oxygen ($O_2$) but an excess of $CO_2$ that is the reason for "bad" or "stale" air and provides the need for air changes.

---

\\Tip:
For light, sedentary activities in homes or offices ventilated through an adequate number of large air distribution outlets, the limit on the average air velocity is often is often recommended as about 0.2 m/s in conditions of little turbulence (approx. 5%). Local zones of discomfort caused by higher values, e.g. in the direct vicinity of air distribution outlets, must also be considered. More precise information is available, e.g. in European standard EN 15251.

\\Note:
Dry air consists of 78.1% nitrogen ($N_2$), 20.93% oxygen ($O_2$), 0.03% carbon dioxide ($CO_2$), and 0.94% argon and other noble gases. Furthermore, "normal air" contains a varying proportion of water vapor and pollutants in the form of nitrogen oxide, sulfur dioxide, waste gases, dust, suspended particles and many kinds of micro-organisms.

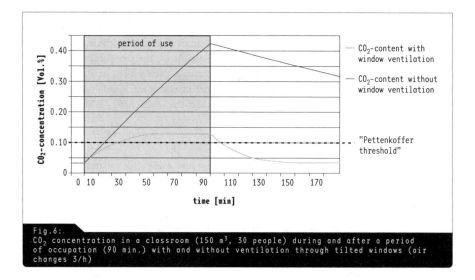

Fig.6:
$CO_2$ concentration in a classroom (150 m³, 30 people) during and after a period of occupation (90 min.) with and without ventilation through tilted windows (air changes 3/h)

> 🗎
Other pollutants

Figure 6 shows the increase in $CO_2$ during the time of occupation of a classroom.

If the air we breathe is contaminated with additional pollutants, e.g. from smoking, emissions from the materials used or from manufacturing processes in the workplace, then there is an additional requirement for fresh air. This requirement is determined by the concentration of pollutants, i.e. the room air must be kept within permissible limits by providing an adequate rate of air changes.

🗎
\\Note:
The "Pettenkoffer threshold" established by Munich doctor and hygienist Max von Pettenkoffer says that the $CO_2$ content of the room air should not exceed 0.10%. This comparatively stringent requirement is used as a comfort criterion for room air quality in most European countries. Adequate means of ventilation must be provided to prevent this value from being exceeded.

🗎
\\Note:
The capacity to adapt has only recently been taken into account in building design. In new regulations, such as European standard EN 15251, the recommended comfort ranges are therefore defined with reference to the coincident outdoor air temperatures. It is expected that other countries will incorporate this new knowledge into their design guidelines.

Tab.1:
Example values of heat given off by people engaged in various levels of activity

| Activity | Total heat given off per person Reference value W |
|---|---|
| Stationary activities performed while sitting, such as reading or writing | 120 |
| Very light physical activities performed while standing or sitting | 150 |
| Light physical activities | 190 |
| Medium heavy to heavy physical activity | more than 270 |

## Intermediate conditions

*User activity* — When engaged in physical effort, the human body expends more energy and needs to give off heat to the environment. › Tab. 1 Less motion expends less energy and the amount of heat given off does not have to be as great. The user's level of activity therefore has just as much influence on his or her perception of comfort. The lower the activity level, the higher the temperature must be to be perceived as comfortable. Moreover, people engaged in little physical activity (e.g. sitting still at a desk) generally tolerate smaller departures from the "ideal temperature" than those with higher activity levels (sport, physical work etc.).

*Clothing* — Independently of the physical activity, clothing influences the amount of heat given off and therefore the body's heat balance. The less clothing is worn, the greater the amount of heat given off by the human body to the air and the surrounding surfaces. Higher temperatures and warm surfaces (e.g. underfloor heating) are often provided in areas where people go barefoot or naked (e.g. in showers or saunas) to avoid discomfort. However, putting on more clothing reduces the body's sensitivity to fluctuations in the ambient temperature because of the clothing's insulation effect.

*Adaptation capacity* — The human body's ability to become used to a specific climatic conditions (e.g. heating periods) in the short, medium or long term, is described as adaptation or acclimatization. This capacity to adapt can therefore mean that initially uncomfortable conditions are perceived as pleasant or at least less distressing over time.

› 🗐

### DETERMINING THE REQUIREMENTS

One of the main requirements of a building is to ensure an appropriate, comfortable room climate in terms of the specified criteria throughout the whole year. While these requirements apply to all the various climate zones, the necessary measures will vary from zone to zone.

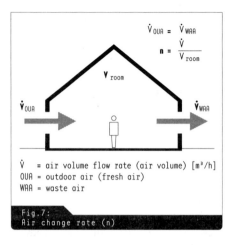

Fig. 7:
Air change rate (n)

### Ventilation requirement

To ensure the air inside a building is of hygienic quality, the room air must be replaced at specified intervals by conducting away the used air and pollutants and replacing them with fresh air. This is generally accomplished by having opening windows in the external walls, but often by mechanical ventilation systems. › Chapter Ventilation systems How often the whole volume of room air must be replaced with fresh air and within what period, e.g. an hour, is given by the air change rate (n). › Fig. 7

How frequently the air in the room must be replaced to ensure healthy room air must be considered above all from the point of view of a room's

> Required minimum air change rate

\\ Note:
Fresh air is air from an unpolluted, natural environment. In ventilation engineering it is different from outdoor air, which is sucked in from outside the building, where the air may be hot and polluted. The $CO_2$ concentration and any airborne pollutants are generally found in outdoor air but the quantities are less than in indoor air. The fresh air demand is normally satisfied by introducing outside air; the term "outdoor air" will therefore be used in the rest of this book.

\\ Example:
Consider a living room with an area of 20 m², a clear ceiling height of 2.50 m, and therefore a volume of 50 m³, occupied by one person. According to Table 3 it would be sufficient to replace half of the air in the room with fresh outdoor air in one hour (n=0.5/h). Table 7 (p. 40) shows for how long, for example, a window has to be open. If the room were occupied by two people, it would require an outdoor air flow rate $\dot{V}$ of 50 m³/h and therefore an air change rate of 1.0/h would be required, i.e. the room air must be completely replaced with outdoor air once per hour.

*Outside air flow rate*

use. The values of air change rate given in specialist literature often differ, and should therefore be taken only as average values for typical room sizes, occupancy densities, and pollution loads. They are particularly useful for preliminary design. › Tab. 2

It is better to calculate the required quantity of fresh outdoor air to be introduced into the room per unit of time, the (outdoor air) flow rate $\dot{V}$ (usually expressed in m³/h) taking into account the expected pollutant load and specific emissions. An air change rate can be calculated from the above flow rate and room volumes, and used as an additional design parameter.

In rooms where the air is not seriously affected by pollutant emissions from building materials or from specific uses, the main factor is the people using the room.

As explained earlier, › Chapter Design principles, Comfort requirements a person's fresh air requirement is met by replacing the room air with air that does not exceed the recommended upper limit for $CO_2$ concentration. The required outdoor air flow rate depends on the type of use, the anticipated pollutant load of the room, and the number of people in the room. In practice, the values in the table are used. › Tab. 3, p. 20

*Different air flow rates*

The air change should, particularly when outdoor temperatures are low, provide only the rate of air changes necessary for hygiene, because an increased supply of outdoor air always results in higher ventilation heat losses. › see below On the other hand, a higher air change rate with high room temperatures can also be used to remove excessive heat.

**Tempering demand**

Tempering demand (heating and cooling energy demand) always arises from a disturbance to the energy balance of a building or room.

Tab. 2:
Recommended air change rates n

| Type of room | n in 1/h |
| --- | --- |
| Living room | 0.6-0.7 |
| Toilets | 2-4 |
| Offices | 4-8 |
| Canteens | 6-8 |
| Bars | 4-12 |
| Cinemas | 4-8 |
| Lecture theatres | 6-8 |
| Committee rooms | 6-12 |
| Department stores | 4-6 |

Tab.3:
Minimum outdoor air flow rate on the basis of DIN EN 15251 (Cat. II)

| Room type | $\dot{V}_{OUA\ min}$ in $m^3$ /(h x person) |
|---|---|
| Living rooms | 25 |
| Cinemas, concert halls, museums, reading rooms, sports halls, retail space | 20 |
| Individual cell offices, auditoriums, classrooms, seminar rooms and conference rooms | 30 |
| Bars | 40 |
| Open-plan offices | 50 |

Energy balance of a room

Room temperatures should remain in a range that is considered comfortable for the room's use. Heat is drawn out of a building by transmission (flow of heat through the building skin because of the temperature differences between indoors and outdoors) and ventilation (exchange of air between indoors and outdoors through uncontrollable leakage points in the building skin and uncontrolled natural or mechanical ventilation). The entry of solar radiation through transparent building components and the heat given off by equipment and persons inside the building brings heat into the building. › Fig. 8

Despite having a building designed for the local climate (e.g. a highly thermally insulated building skin, solar screening etc.), it may still be necessary from time to time, depending on the location of the building, to introduce heat into or remove excessive heat from a building or room artificially (by building services systems), in order to achieve the desired indoor temperature. As this is always linked with an additional energy demand, the objective should be to avoid or keep this additional tempering demand as low as possible by adopting intelligent architecture and building services technology matched to the local climatic conditions. › Chapter Design principles, Covering the demands, Avoidance principle

Heating energy demand

The heat losses of a building or room must therefore be compensated with a corresponding amount of heat from a heating system if the solar and internal gains are insufficient. › Fig. 9 The annual heat energy demand of a building is the sum of the heat losses, taking into account the relevant heat gains, calculated over the year. To be able to assess the thermal quality of a building, it is normally necessary to calculate the annual heat energy demand per square meter of usable space [kWh/($m^2$a)]. This energy parameter allows buildings to be characterized and classified in energy terms in accordance with national energy standards.

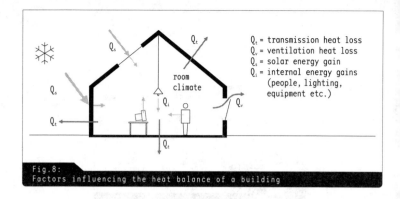

**Fig.8:**
Factors influencing the heat balance of a building

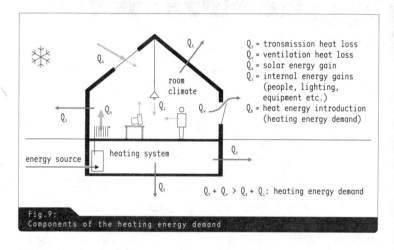

$Q_t + Q_v > Q_s + Q_i$: heating energy demand

**Fig.9:**
Components of the heating energy demand

\\ Note:
The annual heating energy demand provides no information about the actual amount of energy that must be supplied to the building. In order to calculate the end energy demand $Q_e$, which can usually be monitored with an energy meter (gas or electricity meter etc.), the effects of any hot water system connected to the heating system must be taken into account, along with aspects such as the efficiency of the heating system including the relevant losses from heat distribution, and the additional energy required to operate the system (electricity for pumps etc.). To determine the total energy demand including the effects on the environment of a building, the amount of fuel used must be taken into consideration, by calculating either the amount of $CO_2$ emitted or the total primary energy demand (see Chapter Design principles, Covering the demands, Environmental impacts).

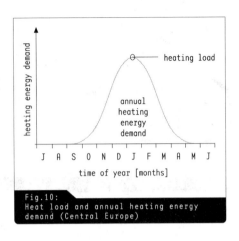

Fig.10:
Heat load and annual heating energy demand (Central Europe)

Tab.4:
Standard indoor temperatures for heated rooms (in accordance with DIN EN 12831 Suppl. 1), where the client has not supplied any other values

| Building or room type | Operative temperature [°C] |
|---|---|
| Living and sleeping rooms | + 20.0 |
| Offices, committee rooms, exhibition rooms, main stairwells, booking halls | + 20.0 |
| Hotel rooms | + 20.0 |
| General retail halls and shops | + 20.0 |
| General classrooms | + 20.0 |
| Theaters and concert halls | + 20.0 |
| WCs | + 20.0 |
| Bath and shower rooms, bathrooms, changing rooms, examination rooms (in general, any use involving undressing) | + 24.0 |
| Heated auxiliary rooms (corridors, stairwells) | + 15.0 |
| Unheated auxiliary rooms (cellars, stairwells, storage rooms) | + 10.0 |

**Heating load**

To design the heating system (heat generation and distribution system) it is necessary to determine the heat load, and hence the maximum required heat output. This value indicates the amount of heat (in W or kW) that must be supplied to a building on the coldest day of the year to compensate for heat losses and achieve the required indoor temperatures. › Tab. 4 and Fig. 10

**Cooling energy demand**

Cooling energy demand arises when the heat gains (as a result of solar radiation entering the building and internal heat loads) are greater than the heat losses (as a result of ventilation and transmission losses)

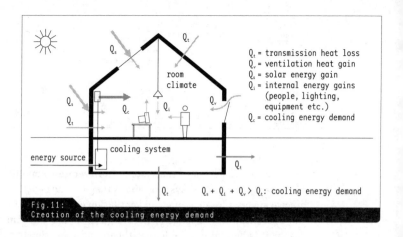

Fig.11:
Creation of the cooling energy demand

and the excess heat cannot be effectively stored in thermal storage masses. › Chapter Design principles, Covering the demands Room temperatures rise and active measures, such as building services systems, cannot prevent the room temperatures exceeding the maximum desired values. › Fig. 11

> ◉ Calculation of the cooling energy demand

The extremely complex physical interactions of the factors involved makes it difficult to predict the anticipated cooling energy demand precisely. In addition to the external and internal loads, the designer must also take into account the processes of heat take-up and emission of the thermal storage mass of the building construction. Dynamic simulation software is normally required to enable precise analysis of a situation in hourly simulation steps in critical or difficult cases.

Cooling load

In normal design situations, therefore, an exact estimate of the cooling energy load is often not made, and a cooling load calculation is performed instead, based on static (constant) conditions. It allows, in particular, an estimate to be made of the maximum cooling load, which can

◉
\\Note:
When outdoor temperatures are high, in addition to the heat gains from the entry of solar radiation and internal heat loads (people, equipment and lighting), the heat gains from ventilation and poorly insulated building components also contribute to the cooling energy demand.

| Tab.5: Examples of internal heat loads | |
|---|---|
| | **Output when operating** |
| Computer with monitor | 150 W |
| Laser printer | 190 W |
| Person, sedentary activity | 120 W |
| Lighting | 10 W/m² of plan area |

be useful in the design of the system. However, this simplified approach often leads to over-sized systems and is therefore no substitute for a more precise analysis by a heating and ventilation engineer.

For the initial selection of a cooling system in a preliminary building design, it is helpful to be able to estimate the approximate heat load of a room and compare it with the capacities of various cooling systems. Table 5 gives some approximate values for common heat loads in an office.

The designer must check the extent to which the demand calculated using a simplified cooling load calculation can also be satisfied from thermal storage masses and ventilation. This usually leads to a more economically efficient size of system.

### COVERING THE DEMANDS
#### Avoidance principle

In order to keep the energy demand of a building as low as possible, in addition to the end energy required for heating, cooling and ventilation, a designer must also take into account user-specific demands (e.g. electrical

\\ Example:
The expected cooling load for an individual cell office with a floor area of 10 m² can be roughly calculated as follows:
According to Table 5, the anticipated internal peak load based on a floor area of 10 m² is 560 W (150+190+120+100) or 56 W/m². However, as loads from devices such as printers or lighting are not constant, these values may be considered a little too high. A reduction of the operating times for printers and lighting to, for example, 50% results in an internal heat load of 41.5 W/m². Solar radiation entering the building through the facade must also be taken into account.

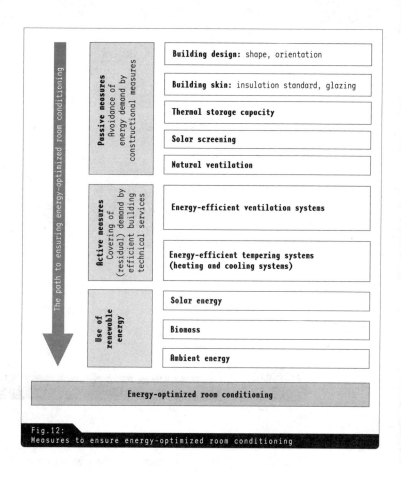

Fig.12:
Measures to ensure energy-optimized room conditioning

consumers such as computers, machinery etc.). A mixture of different passive and active measures may be used to avoid the energy demand in the first place or provide the necessary amount of energy. › Fig. 12

### Passive measures

Passive measures can be used without significant energy cost to influence heat gains and losses. Certainly, the desired room conditions may not always be achieved, but the effects of the passive measures are in the right direction, and active measures further reduce the demand (i.e. the energy demand). There are various options for passive measures, which depend, to some extent, on the fundamental architectural aspects of a building and should therefore be considered at a very early stage in the planning process.

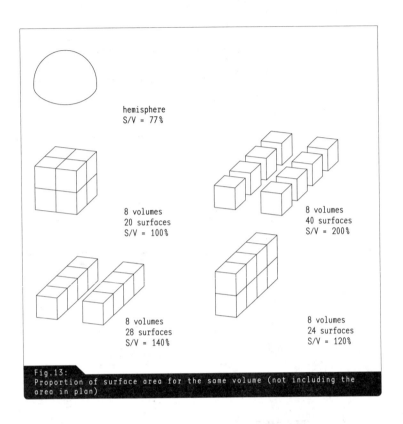

Fig. 13:
Proportion of surface area for the same volume (not including the area in plan)

Building shape

A building interacts with the outside world through its skin. The dimensions of this heat-transmitting thermal skin also determine the amount of heat transferred. The building shape is therefore a very important design parameter for a building's energy balance. The energetic characteristic for the building cubature is called the S/V ratio, which is the ratio of the area of heat-transmitting building skin to the enclosed building volume. › Fig. 13

The various parameters such as room heights, daylight provision in the depth of the room, or functional requirements are normally determined in the design of a building, which then limits the scope of energy optimization to be gained from the building shape. The final building solution often involves a layout with several independent component objects (separate buildings) and particular layouts in plan (end-linked blocks etc.), which leads to comparatively large surface areas with the corresponding energy effects. Compact shapes and the smallest number of separate buildings offer energetically better solutions.

**Alignment and climate zones**

Orientation and climatic zoning within a building can make the most of solar gain to the advantage of energy saving and comfort. In a residential building characterized by a high heating energy demand (e.g. in northern Europe), the rooms with the highest temperature requirements (e.g. living rooms) should be oriented with their largest glazed areas facing the sun so as to maximize the benefits of solar gain. Rooms with lower temperatures (e.g. bedrooms) can face away from the sun. A highly thermally loaded room (e.g. a conference room with a high occupancy rate) should be oriented away from the sun if possible, to avoid additional overheating and any cooling loads from the entry of solar radiation. In addition, zones with unallocated uses and low requirements for temperature and comfort (e.g. circulation areas) can also be placed near the main usage areas to act as buffer zones between the indoor and outdoor climates.

**Insulation standard**

Thermal insulation acts as a barrier between the inside and the outside space of a building. The insulation standard is therefore a measure of the thermal quality of the heat-transmitting building skin. In moderate climatic zones where the outdoor temperatures fluctuate only a little during the course of a day or a year, insulation is usually less important, but there are the occasional exceptions. A good standard of insulation is generally considered useful for avoiding heating and cooling energy demands in hot as much as in cold regions.

Transparent components (windows, skylights etc.) are of special importance here because the insulation standard of a window is usually poorer than that of an opaque component. Depending on the window quality and the building location, the advantages of solar gain through large windows may be cancelled out by increased heat loss in winter.

**Thermal storage**

Every material has some ability to store heat and release it again over time. The amount of heat stored and given out over time depends on the material of the components enclosing the room (walls, floor and ceiling). Concrete or masonry blocks, for example, store more heat than wood or gypsum plasterboard. This is expressed as the "thermal mass" of a room or building. › Fig. 14

Although thermal mass has no direct influence on thermal gains and losses, it is very important to heating and cooling loads and hence the energy demand of a building at any particular time: large thermal masses can store and release more heat from the room air and therefore avoid heating or cooling demand. As a rule, large thermal masses also improve comfort, because surface temperatures rise with increasing air temperatures and take longer to fall with decreasing temperatures.

**Solar screening**

The solar screening of a building is one of the most important passive measures for room conditioning. The amount of heat entering the building, as already indicated, significantly affects the energy demand.

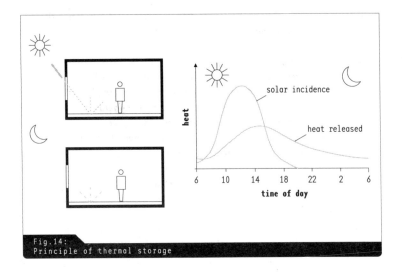

Fig.14:
Principle of thermal storage

Solar screening must therefore be able to fulfill, as far as possible, the opposing goals of maximizing the amount of solar energy entering the room in cold weather and avoiding overheating in hot weather.

Solar screening may be classified as fixed or flexible. Fixed measures such as window sizes, orientation, shading by other buildings (buildings that are part of the project or nearby buildings, canopies, trees etc.) and the overall total solar energy transmittance of the glass used cannot be varied according to the season or the time of day, which results in disadvantages compared with flexible systems. > Fig. 15

\\ Note:
Further information about the design of windows and solar screening elements can be found in *Basics Facade Apertures* by Roland Krippner and Florian Musso, Birkhäuser Verlag, 2008.

\\ Tip:
Depending on use, solar radiation entering the room can cause critical situations at any time of year. This is often the case, for example, for modern office buildings with high internal loads. In combination with a good insulation standard, even in temperate climates (such as in Germany) these buildings have a heating energy demand on very few days of the year. In such cases, for example with the use of large glazed areas, it is particularly important to check that overheating does not occur in summer.

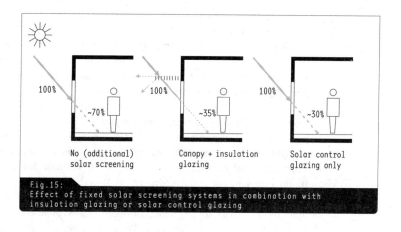

Fig.15:
Effect of fixed solar screening systems in combination with insulation glazing or solar control glazing

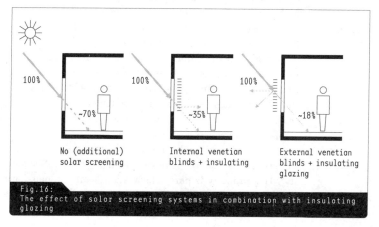

Fig.16:
The effect of solar screening systems in combination with insulating glazing

Flexible measures (Venetian or roller blinds etc.) allow the user short-term influence over how much solar radiation enters the room. In this way the heating energy demand in winter can be reduced as much as possible, while overheating (and the cooling energy demand) is avoided in summer. Flexible systems are therefore preferred to fixed systems. Figure 16 shows the effect of flexible solar screening systems.

Natural ventilation

Natural ventilation is also considered a passive room conditioning measure. As well as satisfying the outside air demand of a building without using energy for ventilation, it is also possible to conduct warm air out of the building and reduce the cooling loads. The special considerations in the design of natural ventilation concepts and the limits of natural ventilation are described in more detail in the chapter on ventilation systems.

### Active measures

The energy demand of a building that cannot be fulfilled by passive measures must be covered actively, by introducing energy. The use of active systems may also bring advantages, although perhaps in other areas. Every concept must therefore be considered as a whole, with all its components and energy sources.

*Efficiency increases*

When considering the provision for the remaining energy demand of a building, it is important that the components used work as efficiently as possible. The designer must ensure that the losses involved in supplying, distributing and transferring energy into the room are as low as possible. There are a multitude of different components available for this, each with its own advantages and disadvantages. However, the components cannot be freely combined with one another. An integrated approach to the design must be used so as not to prevent the development of potential advantages by making poor initial decisions. Figure 17 gives an example of how to increase the efficiency of heating systems.

*Energy transfer*

The selection of the transfer system particularly depends on "radiation proportion" and "controllability".

A heating or cooling body emits its energy into the room by a combination of radiation and convection (the heat is "carried off" by the air). Radiating heat transfer systems are generally advantageous because of the strong influence of room surface temperatures on the operative room temperatures and therefore on people's perception of comfort in a room.

› Chapter Design principles, Comfort requirements

The speed with which a transfer system is able to react to changes in the controls (e.g. by opening or closing a valve) reflects its "controllability". This property is particularly significant if rooms have varying uses or thermal loads.

---

\\ Example:
A conference room is kept at the desired temperature in winter by a heating system. When the conference begins, the room fills with people, all with bodies giving off heat, which means that additional heating is no longer required. In this situation the heating system must be able to reduce its output immediately to avoid overheating and unnecessary energy use.

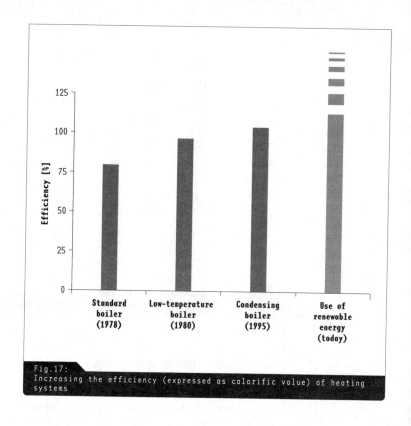

Fig.17:
Increasing the efficiency (expressed as calorific value) of heating systems

Combinations of sluggish radiating surfaces to cover the basic load and quick-reacting air systems for the peaks are therefore often adopted for rooms with frequently changing conditions and high comfort requirements. In these combinations, their disadvantages cancel each other out to produce a powerful, efficient system. Such solutions are associated with relatively high capital costs for their installation. > Chapter Tempering systems, Heat and cold transfer

Distribution

The transport and distribution of energy from the generator to the transfer station is also important for the overall concept. The first thing to consider is that losses occur every time water or air is transported through pipes or ducts. One part of these losses is due to friction on the inside of the pipes, while another is due to temperature losses, so that the temperature of the medium has often dropped by the time it reaches the transfer station. Pipe lengths should therefore generally be kept as short and as well insulated as possible to minimize these losses.

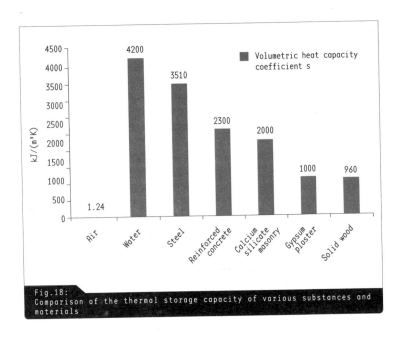

Fig. 18:
Comparison of the thermal storage capacity of various substances and materials

Selecting the transport medium

The selection of the transport medium is of crucial importance. Water and other liquids can store many more times the amount of heat than air. > Fig. 18

When account is taken of the electricity required for the electric fan or pump, this demonstrates that water as a medium for transporting energy is considerably more efficient than air. In terms of the auxiliary energy required for the transport of heat or cold, a water-operated system is preferable to a purely air-operated system. Air-operated systems can be viable, for example, if small heating or cooling loads to be covered are small and a ventilation system has to be used for other reasons. > Chapter Tempering systems, Heat and cold distribution

Energy supply

When deciding on the provision of energy to the building, the question of its source must be considered. In addition to the actual demand (heat, cold, electricity etc.), the decision is influenced above all by availability. With fossil fuels this revolves around the ability to connect to existing services (gas, electricity, district heating etc.); with renewable energy sources the opportunities for their use are important (solar radiation, geothermal, biomass etc.). With some energy sources the ability to store them on site (oil, wood etc.) must be taken into account. > Chapter Tempering systems, Energy supply

As well as the availability of an energy source, the load profile (the changes in energy demand in relation to time) is crucial in the choice of energy source: the demand may vary greatly because of different climatic conditions or uses and the season or time of day. These fluctuations in demand can have a considerable influence on the overall efficiency of a system.

In this context, attention should be paid to the simultaneity of energy supply and requirement and any necessary energy storage measures. A solar energy system for hot service water, for example, works only when there is solar radiation present, i.e. during the day. Domestic hot water demand follows the typical use profile, with the morning and evening peaks being much greater than during the rest of the day. The heat generated during the day must therefore be stored in a buffer, which then makes it available at peak times.

### Fossil and renewable energy sources

While fresh air provided by a mechanical ventilation system generally uses electricity as the energy source, there is often a choice of energy sources for generating heat and cold, all of which can be used in a variety of generation systems. Fossil fuels and renewable energy sources are considered separately below.

Fossil fuels

Fossil fuels (e.g. oil, gas and coal) were created over a long period of time though biological and physical processes below and on the surface of the Earth and therefore cannot be recreated in the short term. The Earth's existing supply cannot be renewed; the stocks are finite. These energy sources are based on carbon, which is released into the atmosphere as $CO_2$ by combustion and is a significant cause of global warming.

In the past, almost all buildings obtained their energy from fossil fuels, and the necessary technology is therefore widely developed and can operate at a high level of efficiency. In spite of these positive developments, the energy supply to buildings is still a major source of global $CO_2$ emissions, and calls for increased use of renewable energy.

Renewable energy sources

>

Renewable energy sources are inexhaustible on a human scale and therefore sustainable: they can be exploited without permanently damaging the environment.

\\Note:
The term "sustainable" comes from forestry and describes the principle of only cutting down as many trees in one area as will be replaced in the same year. Expressed more generally, the term refers to using a natural system only in ways that preserve its important characteristics over the long term.

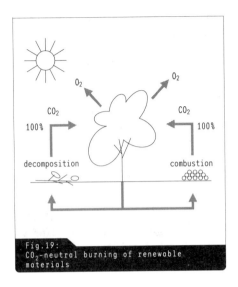

Fig. 19:
$CO_2$-neutral burning of renewable materials

In this sense, solar energy (electricity or heat generation), hydropower, wind, geothermal and bioenergy (biomass such as wood, or biogas such as methane from sewage treatment plants) are considered to be renewable energy sources. Although the burning of biomass and biogas also releases $CO_2$ into the atmosphere through combustion, this carbon dioxide was taken up in the recent past from the atmosphere as the plants grew and would have been released again by the plants during the natural process of rotting. The burning of biological energy sources is therefore described as $CO_2$-neutral. ❯ Fig. 19

Thus, it is not the release of $CO_2$ in general that is critical, but rather the avoidance of $CO_2$ emissions that would not have occurred and polluted the atmosphere without the combustion process.

The technological systems for using renewable energy in buildings have recently undergone considerable advances, and years of experience have proved them reliable.

### Environmental impacts

For a long time, consideration of the energy demand of a building was limited to the demand (e.g. heating energy) arising in the building. To assess the effects on the environment, this way of looking at the situation is not enough, because the equipment losses in the building (e.g. losses in the boiler during the heating of the water and in its transport from the boiler to the radiators) are not taken into account. For another thing,

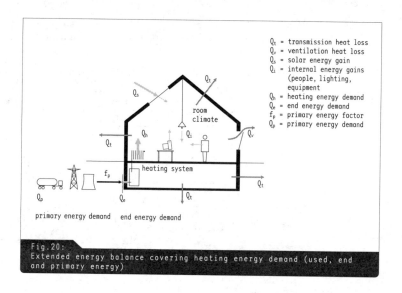

Fig. 20:
Extended energy balance covering heating energy demand (used, end and primary energy)

energy is lost en route from the point of its generation to the point of supply at the building. We therefore now distinguish between the energy use in the building, the end energy demand at the building boundary, and the primary energy demand, which describes the demand on naturally created energy sources. › Fig. 20

Primary energy factors

The required expenditure for the end energy demand including the pre-chains (material conditions and auxiliary energy) for the extraction, processing, conversion, transport and distribution of an energy source is defined using the primary energy factor. › Tab. 6

Electricity is normally produced in several ways, e.g. in coal, hydroelectric or nuclear power stations. A primary energy factor for the "electricity mix" can be calculated from the proportion of fossil fuels, nuclear and renewable energy sources used in the generation of electricity. This factor varies from country to country and can therefore be used to judge the effect of the use of that electricity.

$CO_2$ emission factors

g/kWh$_{end}$

In a similar way to the primary energy factors, $CO_2$ emission factors can be calculated to estimate the amount of greenhouse gases emitted (in grams) per kWh of end energy used. The unit of measurement for this factor is [g/kWh$_{end}$]. In addition to $CO_2$ the factor also takes into account the amount of other pollutants emitted and summarizes the resulting greenhouse effect as a "$CO_2$ equivalent". › Tab. 6 By multiplying this factor by the energy demand of a building, the effect of the energy supplied to the building on global warming can be calculated.

Tab.6:
Example of primary energy and $CO_2$ emission factors in Europe (based on DIN-V-18599-1:2007-02)

| Energy source | | Primary energy factor (non-renewable proportion) [$kWh_{prim}/kWh_{end}$] | $CO_2$-emission factor ($CO_2$-equivalent) [$g/kWh_{end}$] |
|---|---|---|---|
| Fuel | Oil EL | 1.1 | 303 |
| | Gas H | 1.1 | 249 |
| | Liquid gas | 1.1 | 263 |
| | Coal | 1.1 | 439 |
| | Brown coal | 1.2 | 452 |
| | Wood | 0.2 | 42 |
| District heating (at 70%) from power-heat coupling | Fossil fuels | 0.7 | 217 |
| | Renewable fuels | 0.0 | |
| District heating from heating plants | Fossil fuels | 1.3 | 408 |
| | Renewable fuels | 0.1 | |
| Electricity | Electricity mix | 2.7 | 647 |
| Environmental energy | Solar energy, ambient heat | 0.0 | |

Figure 21 gives an example of the effects of the choice of energy source on the primary energy demand and the $CO_2$ equivalent emissions of a building.

As can be seen in Figure 21, if electricity is used to provide for an end energy demand of 50 kWh per square meter of residential space per year, almost 3 times the amount of primary energy is required compared

\\ Note:
The unit of measurement of the primary energy factor is [$kWh_{prim}/kWh_{end}$]. The factor indicates how much primary energy in kWh (for example, the quantity of a particular energy source, e.g. coal) is required for the provision of one kWh of end energy (e.g. electricity or heat energy.)

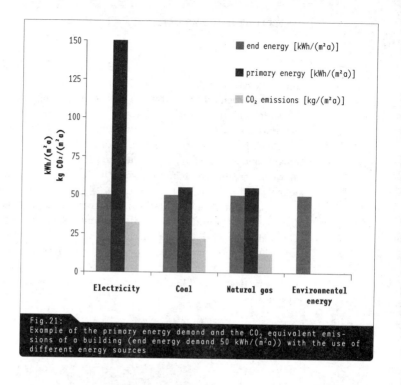

Fig. 21:
Example of the primary energy demand and the $CO_2$ equivalent emissions of a building (end energy demand 50 kWh/(m²a)) with the use of different energy sources

with covering the demand using environmental energy. With gas or coal, the primary energy used is only 1.1 times, just 10% more. Environmental energy, on the other hand, is almost primary energy neutral. The same applies to $CO_2$ emissions.

As far as local circumstances allow, the use of electricity and fossil fuels should therefore be avoided.

## VENTILATION SYSTEMS

As can be seen from the discussions of comfort requirements, factors such as air temperature, air velocity, humidity, cleanliness and composition are particularly important in the perception of comfort in a room. These factors are strongly influenced by ventilation, and the particular ventilation system used in the building plays a significant role.

The main task of ventilation is to take odors, water vapor, carbon dioxide and air with a high concentration of pollutants out of a room and, above all, create and maintain a good, uniform thermal environment in a room.

The heating and ventilation industry makes a fundamental distinction between <u>natural (free) ventilation</u> and <u>mechanical ventilation</u>. The boundaries between free and mechanical ventilation (e.g. ventilation systems, heating and ventilation systems, often referred to as HV systems, and air conditioning systems) are fluid and not always clear in speech or in practice. Figure 22 illustrates one possible division for ventilation systems.

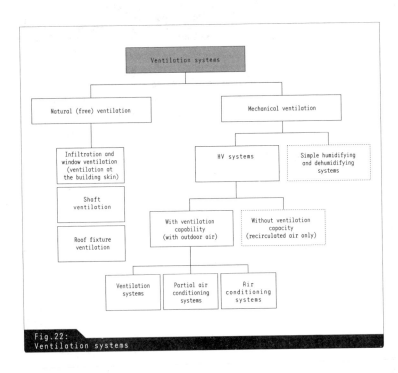

Fig.22:
Ventilation systems

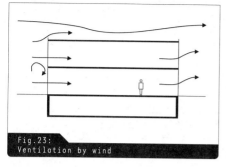
Fig. 23:
Ventilation by wind

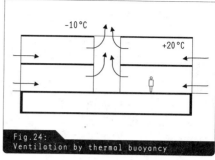
Fig. 24:
Ventilation by thermal buoyancy

## NATURAL VENTILATION

With natural (free) ventilation, the movement of the air in a room is caused exclusively by pressure differences at the surface and inside the building. These pressure differences are created by wind › Fig. 23 or temperature differences (thermal buoyancy). › Fig. 24

As shown in Figure 22 natural ventilation can be divided into three categories:

_ Infiltration and window ventilation (ventilation at the building skin)
_ Shaft ventilation
_ Roof ventilation

The required air volume flow depends greatly on the weather, room temperature, and the arrangement and aerodynamic design of the ventilation openings.

### Infiltration ventilation

The term "infiltration ventilation" describes the exchange of air that takes place through leaks in the building, mainly through gaps around opening windows and doors. This form of ventilation it is associated with a number of problems.

In conditions of still wind and small temperature differences, the provision of the minimum number of air changes required for hygiene cannot be guaranteed. Moreover, infiltration ventilation cannot normally be influenced by the user, and continuous uncontrolled ventilation results in increased heat loss, which may lead to serious building damage. To avoid this and allow a rate of air change to match the demand, the external skin of an energy-saving building must be as airtight as possible (including the windows), and have no uncontrolled gaps or joints. The outside air demand must also be covered by other means.

Tab. 7:
Air change rates using window ventilation

| Type of window ventilation | Air changes |
|---|---|
| Windows, doors closed (infiltration ventilation only) | 0 to 0.5/h |
| One-sided ventilation, windows tilted, no slatted roller blinds | 0.8 to 4.0/h |
| One-sided ventilation, window half-open | 5 to 10/h |
| One-sided ventilation, window fully open[1] (purge ventilation) | 9 to 15/h |
| Cross ventilation (purge ventilation by opposing windows and doors) | to 45/h |

[1] Purge ventilation for 4 minutes results in one air change.

### Window ventilation

As a rule, windows or other regulated openings (e.g. flaps) provide the natural ventilation of buildings or rooms, and by necessity are either kept open for brief periods (purge ventilation) or kept open over a longer period (permanent ventilation). For most buildings this is the way a room is ventilated to keep it comfortable for most of the year.

In winter and high summer, depending on the climate zone, the high heat losses or heat loads associated with window ventilation create problems and make it suitable for short-term purge ventilation only.

*Window types*  The air entry and exit openings of sliding and pivot-hung windows are equally large and adjustable, which makes them more suitable for ventilation than tilt windows. › Fig. 25

*Air flow behavior*  The air flow behavior for window ventilation is different in winter and summer. This difference depends on the temperature difference between indoors and outdoors. › Fig. 26

*Cross ventilation*  Opening windows in just one wall will only provide one-sided ventilation. Ideally, a building should be ventilated by cross ventilation from windows on opposite sides. In residential buildings, there should be adequate cross ventilation, or at least ventilation across a corner. › Fig. 27

*Air change rate*  Table 7 gives rough guidance values for air change rates that various window ventilation arrangements may achieve.

The expected rates of air change provided by window or infiltration ventilation fluctuate and depend strongly on the wind speed and building geometry. There are limits to the size of rooms or buildings that can be ventilated naturally. For the one-sided ventilation of a room with a clear ceiling height of up to 4 m, the room depth should be not more than 2.5 times the room height. With cross ventilation this ratio rises to 5. › Fig. 28

Table 8 provides an overview of the guidance values for the design of natural ventilation systems.

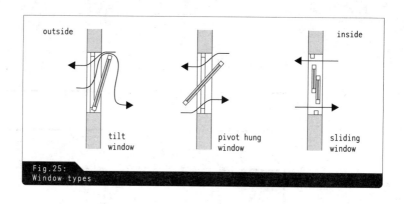

Fig.25:
Window types

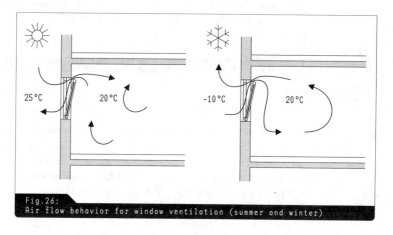

Fig.26:
Air flow behavior for window ventilation (summer and winter)

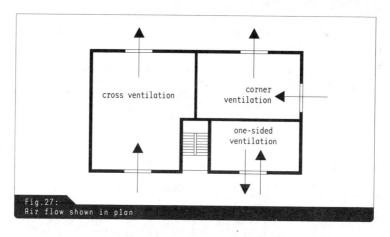

Fig.27:
Air flow shown in plan

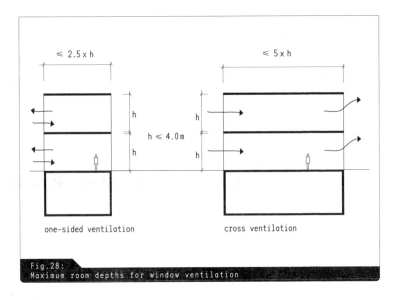

Fig.28:
Maximum room depths for window ventilation

User behavior

The rate of air change achievable through window ventilation depends greatly on user behavior. It is certainly an advantage for the user to be able to control the air flow into a room directly, but in practice rooms are usually ventilated too much or too little. Controlled air exchange cannot take place by user-controlled window ventilation.

Tab.8:
Guidance values for room depths and ventilation cross sections (in accordance with German workplace regulations)

| System | Clear ceiling height (h) | Room depth maximum | Inlet and outlet air cross-sections in $cm^2$ per $m^2$ floor area |
|---|---|---|---|
| One-sided ventilation | up to 4.0 m | 2.5 x h | 200 |
| Cross ventilation | up to 4.0 m | 5.0 x h | 120 |
| Cross ventilation with roof fixtures and openings in an external wall or in two opposing outside walls | more than 4.0 m | 5.0 x h | 80 |

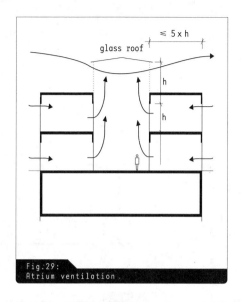

Fig. 29:
Atrium ventilation

## Shaft ventilation

A further form of natural ventilation is shaft ventilation. It is based on the principle of thermal buoyancy, › Fig. 24, p. 39 which involves a drop in temperature between the indoor air and the outdoor air at the top of the shaft, and the suction effect of the wind passing over the top of the shaft.
› Fig. 30

If these conditions are not present, e.g. in summer when the outdoor and indoor air temperatures may be equal or when there is no wind, this

\\ Tip:
Atria should be designed so that rooms can be provided with light and air, even in deep buildings. Natural ventilation works in the same way as a chimney: the heated air rises and escapes through openings in the roof. This creates low-pressure zones, which suck waste air out of the adjacent rooms. If the atrium is covered with glass, sufficient vertical exhaust air openings must be provided to prevent hot air building up in summer (see Fig. 29).

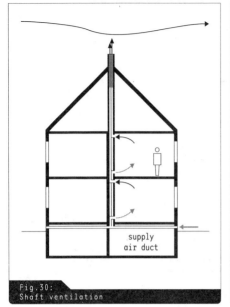

Fig. 30:
Shaft ventilation

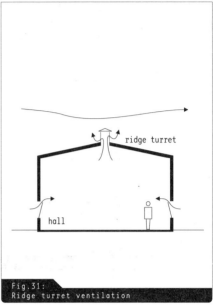

Fig. 31:
Ridge turret ventilation

type of ventilation is rendered ineffective without a fan. Shaft ventilation is therefore only suitable for rooms in which there is a primary need for water vapor to be conducted away, such as bathrooms and kitchens. This can be easily accepted in the short term with plastered surfaces, and the shaft effect will resume when there is enough buoyancy again. It is important to ensure that each ventilated room has its own shaft, so that sufficient air can flow through it at all times, and that odors from other rooms can be excluded.

**Roof fixture ventilation**

Roof fixture ventilation is natural ventilation that occurs through fixtures such as ridge turrets, roof lanterns and similar exhaust air openings in the roofs of buildings. › Fig. 31

This form of ventilation, like shaft ventilation, is primarily based on the principle of thermal buoyancy caused by temperature differences between the air indoors and outdoors. Roof fixture ventilation is mostly used for rooms with ceiling clearances in excess of approximately 4 m, › Tab. 8 i.e. large halls (industrial sheds). Roof ventilation has no associated operating costs but is responsible for high heat losses in winter. Nowadays, mechanical fans with heat recovery are used.

## MECHANICAL VENTILATION

Mechanical ventilation systems include <u>simple ventilation systems</u> (e.g. shaft ventilation systems with fans or external wall ventilators) and <u>heating and ventilation (HV) systems</u>, which prepare air in a central mechanical equipment room and supply it to the rooms through air distribution systems (air ducts, shafts). › Fig. 22, p. 38

### Heating and ventilation systems

Heating and ventilation systems may be <u>with and without ventilation</u>. HV systems without ventilation components are systems that only recirculate air, without adding fresh air. These systems are mainly used in certain industrial manufacturing processes, and are not considered further here.

HV systems with ventilation components

HV systems that ventilate are intended to renew and filter the room air. They always introduce a certain proportion of outdoor air and have an exhaust or waste air outlet so that the room air can be constantly renewed. If the indoor air is not contaminated with odors or pollutants, part of the exhaust air can be recirculated through a mixing chamber, where it is mixed with outdoor air. The HV systems also treat the air thermodynamically by heating, cooling, humidifying and dehumidifying it. Thus, the systems can be differentiated by their treatment of the supply air:

_ Ventilation systems with no thermodynamic air treatment capability, or only one (e.g. heating only)
_ Partial air conditioning systems with two or three thermodynamic air treatment capabilities (e.g. heating and cooling, or heating, cooling and dehumidifying)
_ Air-conditioning systems with all four thermodynamic air treatment capabilities (heating, cooling, humidifying and dehumidifying)

Centralized HV systems

Centralized HV systems always require machinery to process the air and, above a certain size, must be installed in their own rooms (HV

\\Note:
Heat recovery equipment extracts heat energy from the exhaust air and introduces it back into the supply of outdoor air. It saves valuable heating and cooling energy, which otherwise would have been necessary to temper the supply air.

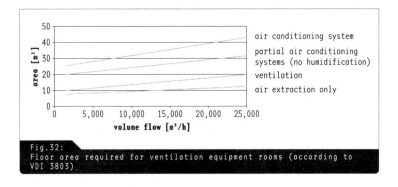

Fig. 32:
Floor area required for ventilation equipment rooms (according to VDI 3803)

equipment rooms). The requirement for outdoor air must first be calculated from the anticipated use, outdoor climate and the desired indoor climate. › Chapter Design principles, Determining the requirements

If the required (outdoor) air volume flow $\dot{V}$ is known, then by taking into account the type of air processing required, it is possible to determine the amount of space a centralized HV system will require using the conservative values in Figure 32.

### Distribution

If air conditioning is performed centrally, the air has to be distributed throughout the building through ducts, vertically in shafts and horizontally in the floor zone. The design should make allowance for the space required.

› ⓘ
Shape and space requirement of ventilation ducts

The shape of a ventilation duct is determined by the flow characteristics and the available space. Circular cross-sections are inexpensive and

ⓘ
\\ Note:
An alternative to a centralized air-conditioning system is a decentralized ventilation system, which can minimize the space required for shafts and ducts. By using decentralized ventilation units, every room can be supplied directly with the required outdoor air, and the supply air can also be tempered to individual needs by heating or cooling (see Chapter Tempering systems, Heat and cold distribution).

ⓘ
\\ Note:
A ventilation system handling supply and exhaust air requires separate supply and exhaust air ducts. This must be taken into account when allocating space. Supply and exhaust air ducts should not cross one another, and suspended ceilings must be able to accommodate two ducts, one on top of the other, otherwise the ceiling height may be considerably reduced in certain circumstances. Where minimum ceiling heights must be observed this can considerably increase building costs.

> \\ Example:
> Calculation of the required supply air duct cross-section to supply a room with an air volume of 1200 m³/h at an air velocity of 3 m/s in the duct:
>
> $A = \dot{V}/(v \times 3600)$
> $= 1200 \text{ m}^3/\text{h} / (3 \text{ m/s} \times 3600 \text{ s/h})$
> $= 0.1111 \text{ m}^2 = 1111.11 \text{ cm}^2$
>
> If a rectangular supply air duct with a space-saving side ratio of 1:3 is to be used, the cross-section can be calculated as follows:
>
> $b/h = 1/3$
> $b = 3 \times h$
> $A = b \times h = 3 \times h^2$
> $h = \sqrt{(A/3)}$
> $= \sqrt{(1111.11 \text{ cm}^2/3)}$
> $= 19.24 \text{ cm} = \underline{20 \text{ cm}}$ selected height of duct.
> $b = 3 \times h$
> $= 3 \times 20 \text{ cm} = \underline{60 \text{ cm}}$
>
> If a circular duct cross-section had been chosen, it would have had a diameter of about 40 cm and a correspondingly greater space requirement (see Fig. 33).

have good flow characteristics but need more space than rectangular ones, which are therefore used more frequently and are available in a side ratio of 1:1 (square) to a maximum of 1:10 (flat rectangle). Abrupt changes in the direction of the ducts should be avoided in order to minimize the pressure loss and the accompanying air noise. Heavy ducts and ducts with large cross-sections (and hence lower air speeds) also reduce air noise.

In supply air ducts, the air is commonly transported at a velocity of v = 3–5 m/s (but in residential buildings at velocities of v = 1.5 m/s and often less). From the above assumptions and the required air volumes, a conservative value for the required cross-section of a duct can be calculated from the following formula:

$$A = \frac{\dot{V}}{v \times 3600} \; [\text{m}^2]$$

> Insulation

in which: A: Duct cross-section in m²; $\dot{V}$: Air volume flow rate (air volume) in m²/h; air velocity in the duct in m/s (= 3600 × m/h).

Ventilation ducts must be insulated to reduce noise, to minimize heat losses when being used for heating, and to avoid condensation. Approximately 5 cm of insulation is generally used. This must be added to the duct dimensions when considering the space requirements. Added to this is another 5–10 cm working space around the duct. > Fig. 33

> Fire safety and sound insulation

In addition, the ventilation ducts must incorporate special measures to ensure fire safety and provide sound insulation because ventilation ducts often cross fire compartments and usage zones. Suitable fire safety flaps or sound dampers with larger external dimensions than the ducts themselves must be designed into the system and remain accessible for maintenance.

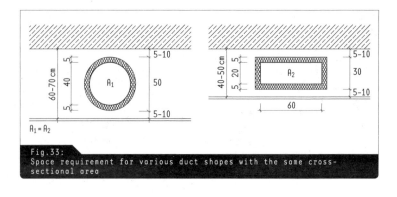

Fig. 33:
Space requirement for various duct shapes with the same cross-sectional area

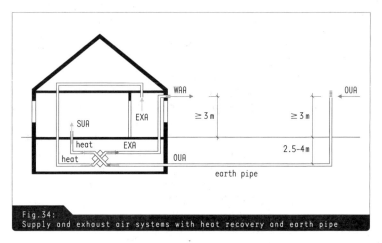

Fig. 34:
Supply and exhaust air systems with heat recovery and earth pipe

Waste and outdoor air ducts

Outdoor air intakes and waste air outlets must have protection against the entry of rain, birds and insects, and be at least 3 m above ground level. › Fig. 34 Taking air into the building through an earth pipe approximately 2.5–4 m underground is an effective means of saving energy. The air is precooled in summer and prewarmed in winter by the relatively constant temperature of the soil. › Chapter Tempering systems, Energy supply

Heat recovery

The heat present in the exhaust air can be recovered and used to warm the outside air in a heat exchanger. The streams of exhaust and cold supply air cross one another in the heat exchanger without mixing, so that there is no transfer of pollutants. The efficiency can be up to 90%, depending on the type of heat exchanger. › Fig. 34

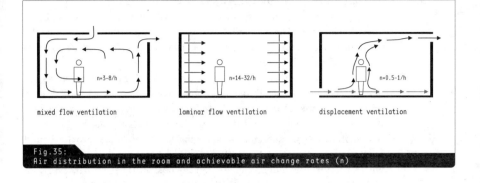

Fig. 35:
Air distribution in the room and achievable air change rates (n)

### Air distribution in the room

The method of distributing the conditioned air is crucial to the perception of comfort in a room. There are three basic ways of distributing air in a room: › Fig. 35

_ Mixed flow ventilation
_ Laminar flow ventilation
_ Displacement ventilation

Mixed flow ventilation

Mixed flow ventilation is the most common means of introducing air into a room. Supply air is blown in at the ceiling or on the walls at a relatively high velocity, and mixes with the still room air.

Laminar flow ventilation

Laminar flow ventilation is a method of ventilating a room used for special purposes. The supply air is introduced over the whole surface of wall or ceiling and removed from the opposite side. Typical uses are operating theaters and clean rooms, where the method of ventilation must ensure that the supply air and room air do not mix in order to create an extremely clean environment.

Displacement ventilation

Displacement ventilation is widely accepted as a particularly energy-saving and comfortable form of ventilation. Air some 2–3 K cooler than the room air is introduced into the room close to the floor at a slow velocity (< 0.2 m/s). The air is distributed at floor level in the room and forms a pool of fresh air. The sources of heat in the room, such as people or computers, cause the fresh air to rise by convection, and therefore supplies everyone with adequate fresh air. The air change rate can thus be reduced to the minimum required for hygiene (n = 0.5–1.0/h) and the energy demand lowered. Displacement ventilation works independently of the room depth and volume of a building and can therefore be used for very deep buildings and halls with a large air demand and low cooling load (up to about 35 W/m$^2$), e.g. in theaters, sports halls, or offices.

## DECIDING ON THE RIGHT SYSTEM

Every ventilation system has its own series of advantages and disadvantages. › **Tab. 9** The principle is to choose a system that can provide the right amount of fresh air to the building or room as efficiently as possible in terms of energy and with a high degree of comfort.

If possible, preference should be given to natural ventilation, as the use of mechanical ventilation is almost always associated with higher building and operating costs. Also to be borne in mind is the great space requirement for the equipment, the air ducts within the building or room to distribute the air, and measures to ensure fire safety and provide sound insulation. On the other hand, there is frequently a higher quality of use with a comfortable indoor climate and the possibility of heat recovery.

Reasons for the use of mechanical ventilation

Mechanical ventilation should be installed only if it is necessary for functional (constructional or user-related) reasons, or if overall savings in energy can be expected:

- Windowless or internal rooms require a supply of outdoor air.
- Installation is advised for buildings taller than about 40 m. Wind pressures and convection can cause severe draughts when windows

Tab. 9:
Characteristics of ventilation systems

| | Advantages | Disadvantages |
|---|---|---|
| **Natural ventilation** | No energy required to drive the system or to condition the air | Effectiveness depends on the climatic conditions (wind velocity and temperatures) |
| | Reduction of space taken up by installations (no air ducts, equipment rooms etc.) | Functioning depends on the building structure and room depths |
| | Lower initial investment and maintenance costs | High heat losses in winter |
| | Optimum relationship with outdoor world (with window ventilation) | Heat recovery impossible or very difficult to implement |
| | High user acceptance | |
| **Mechanical ventilation** | Easily regulated | High initial investment, operating and maintenance costs |
| | Heat recovery possible | Increased space requirement for equipment and air ducts |
| | Suitable for all thermodynamic air treatment functions (heating, cooling, humidifying and dehumidifying) | Low user acceptance, in particular due to lack of user influence |
| | Filters can be incorporated where the air is contaminated | |

are opened in buildings of this height. This would make the upper stories almost impossible to use without special constructional measures such as double facades to counter this effect.

- Buildings in locations with strong odor or noise loads or waste gas emissions make its installation worthwhile.
- Very deep rooms in which free ventilation is not sufficient to provide adequate air exchange require mechanical ventilation. ˃ Fig. 28, p. 42
- In theaters, cinemas and other places of assembly, the absence of windows, or a relatively small window area – linked with high occupancy rates – makes natural ventilation unviable.
- Rooms with a specified indoor air quality in terms of microbial levels, temperature, humidity etc., e.g. operating theaters, museums or special production facilities (microprocessors etc.), make mechanical ventilation necessary.
- Rooms with high thermal loads (e.g. computer centers), where cooling is required, make its installation worthwhile.
- As mentioned above, mechanical ventilation can be an effective means of saving energy. Ventilation systems, in particular those with heat recovery, reduce ventilation heat loss and are an indispensable part of the passive house energy concept.

## TEMPERING SYSTEMS

The purpose of an active tempering system is to use an energy source as efficiently as possible to provide any additional heat or cold necessary to create a comfortable room climate and introduce it into a room. › **Chapter Design principles, Comfort requirements**

In general terms, a tempering system might consist of an energy source, a technical system for creating heat or cold, some method of storage if necessary, and a means of distribution and delivery – in combination with user need-related controls and the object to be tempered (building or room). › Fig. 36

All elements of an active tempering system must be fine-tuned to match one another so that the system can cover the demand › **Chapter Design principles, Determining the requirements** of the building in an efficient manner at all times.

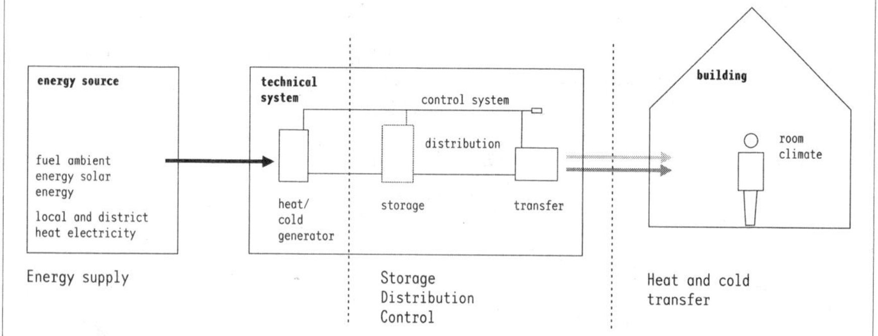

› ◍

| energy source | technical system | | building |
|---|---|---|---|
| fuel ambient energy solar energy | control system | | room climate |
| | distribution | | |
| local and district heat electricity | heat/cold generator | storage transfer | |
| Energy supply | Storage Distribution Control | | Heat and cold transfer |

Fig. 36:
Active tempering system

◍

\\Note:
The pipe that carries the water from the heat generator to the transfer system is known as the "feed", and the one carrying the water back, the "return". The temperatures of the water in these pipes are important parameters for the ability to combine generators of heat or cold with transfer systems.

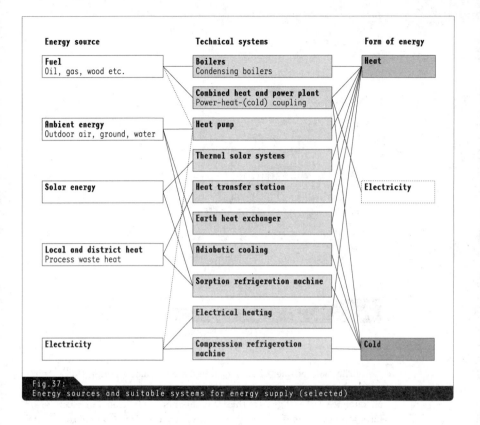

Fig.37:
Energy sources and suitable systems for energy supply (selected)

### ENERGY SUPPLY

The supply of heat and cold in the building depends on the energy sources available. Figure 37 shows possible ways of combining various energy sources with selected systems for supplying energy.

### Fuels

The use of fossil fuels such as oil, natural gas or coal is mainly concerned with generating heat in room conditioning.

The burning of fossil fuels is criticized because of the environmental effects (primary energy and $CO_2$ emissions). $CO_2$-neutral fuels should be used in preference. › Chapter Design principles, Covering the demands

Boilers

All over the world, heat is supplied to buildings primarily by burning fuels in central heating boilers. The heat released during this process is given up in a heat exchanger into a heat medium, usually water, and distributed in the building (central heating).

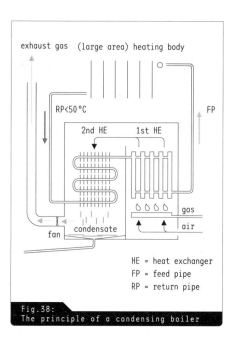

Fig. 38:
The principle of a condensing boiler

Condensing boilers

The most efficient boiler is called a condensing boiler; this extracts additional heat from the exhaust gas and through this can achieve high efficiencies. Oil and gas are the most common fuels used, but recent years have seen systems that can operate with wood pellets (combustible material made of compressed wood in the shape of small rods) as their fuel. > Fig. 38 Condensing boilers are best combined with heating radiators because they require low return water temperatures.

\\Tip:
The low system temperatures of condensing boilers mean they also work well in combination with thermal solar systems. The low exhaust gas temperature requires the chimney for a condensing boiler to be lined with a condensation-resistant inner duct and equipped with a fan because the effect of convection is too weak.

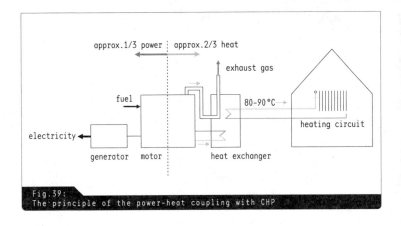

Fig. 39:
The principle of the power-heat coupling with CHP

**Power-heat coupling/ combined heat and power plants**

In addition to burning fossil or biological fuels to generate heat, buildings can also be heated by the heat generated as part of other processes. One important example is the use of the heat that arises during the generation of electricity.

Like heat generation, electricity generation usually relies on combustion processes that produce hot exhaust gases. Two forms of energy are produced from transferring this heat from the combustion process, e.g. in a water-operated system: electricity (power) and heat. These processes are therefore referred to as "power-heat coupling" (PHC). In terms of the amount of fuel used, these processes are very efficient, because most of the energy contained by the fuel can be extracted and used.

The PHC principle can be used equally well in a large heating power station as in a medium-sized plant (e.g. supplying a residential area), or a small one (e.g. supplying a single building). The medium or small electricity and heat plants are called combined heat and power plants (CHP) and supply heat mainly in the form of hot water at temperatures of 80–90 °C. › Fig. 39 The fuels used are most often natural gas and light heating oil, but biogas and biofuels (e.g. rapeseed oil) are also used.

The heat produced by the PHC plants can be used like heat generated from solar energy › see below: Solar energy with the help of sorption refrigeration machines for cooling. These processes are therefore referred to as "power-heat-cold coupling" (PHCC) systems.

**Ambient energy**

There are various possibilities for using the energy potential of a building's surroundings for cooling or heating. In addition to the temperature level of the outdoor air or the room exhaust air there are the relatively

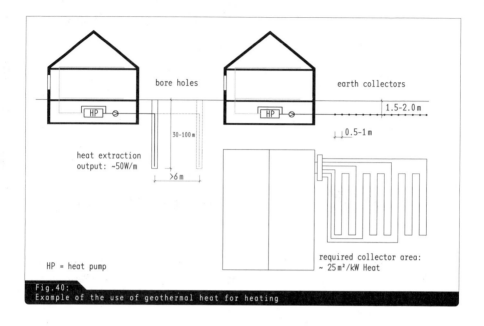

Fig.40:
Example of the use of geothermal heat for heating

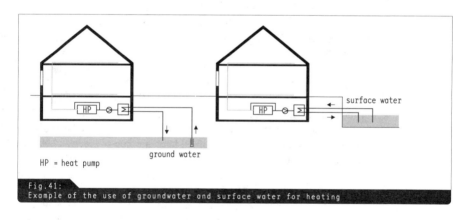

Fig.41:
Example of the use of groundwater and surface water for heating

constant temperatures at various depths of the soil, the groundwater or nearby bodies of water. › Figs 40 and 41

The energy obtained from these sources can be used in winter for heating and in summer for cooling buildings. To be used for heating, the temperature level usually has to be raised by introducing additional energy.

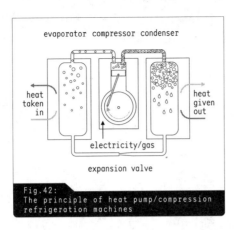

Fig.42:
The principle of heat pump/compression refrigeration machines

Heat pumps

This is done using heat pumps, which work on the same principle as refrigerators. > Fig. 42 The ambient energy source must be augmented by electricity or gas in order to raise the temperature to the desired level and operate the system efficiently.

The use of gas offers primary energy advantages compared with electricity. > Chapter Design principles, Covering the demands

Earth heat exchangers

The use of the heat of the earth ("geothermal energy") is mainly by earth collectors or bore holes in combination with heat pumps. The geothermal heat can also be used simply for preconditioning outdoor air, with the air being fed through an air-earth heat exchanger before it enters the building. In winter the air is prewarmed, and in summer the warm outdoor air is precooled before it enters the building, thus contributing to a reduction of the heating and cooling energy demand. > Fig. 43

\\Tip:
The higher the temperature of the heat source, or the lower the temperature difference between the heat source and heating circuit, the more efficiently the heat pump works. The use of heat pumps to generate heat is most suitable if the system temperatures are low and the heating surfaces large, such as in underfloor heating.

\\Note:
The additional resistance to flow caused by the air having to pass through the pipes of an earth heat exchanger makes mechanical ventilation plant necessary.

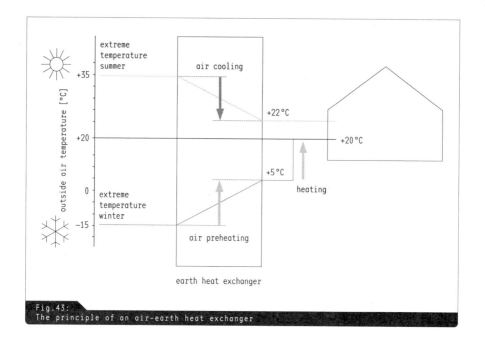

Fig. 43:
The principle of an air-earth heat exchanger

Adiabatic cooling

In addition to underground and surface water as sources of natural energy for heating and cooling as described above, water can also be used directly for cooling buildings. Just like the formation of sweat on the human skin, the evaporation of water draws energy out of the air and cools it. This effect can be created directly with buildings and rooms or indirectly by evaporation cooling, also known as adiabatic cooling. With direct adiabatic cooling the supply air is either passed over open water surfaces or plants or, particularly with mechanical ventilation systems, water is sprayed into the air so finely that it does not reach the floor as it drops but remains in the air as water vapor and thus cools the supply air. A disadvantage of this form of adiabatic cooling is that the supply air becomes increasingly humid. › Chapter Design principles, Comfort requirements, Thermal comfort This makes direct adiabatic cooling most useful in hot, dry climate zones.

By combining the evaporation principle with a heat exchanger, it is possible in mechanical ventilation systems to produce a temperature reduction without increasing the humidity. This is often called indirect adiabatic cooling. It allows the exhaust air from a room to be cooled by evaporation and then expelled from the building through a rotating heat exchanger. The heat exchanger transfers the cold energy highly efficiently into the counter-rotating flow of the supply air. › Fig. 44

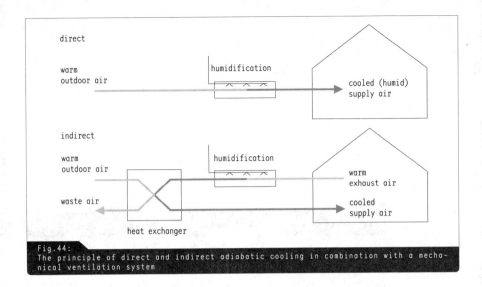

Fig.44:
The principle of direct and indirect adiabatic cooling in combination with a mechanical ventilation system

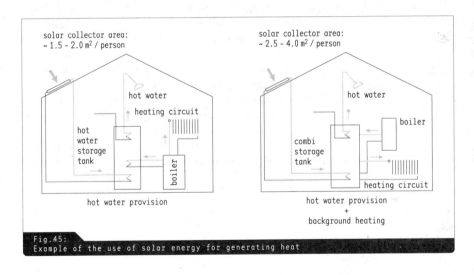

Fig.45:
Example of the use of solar energy for generating heat

### Solar energy

Solar energy systems convert sunlight into heat, which is then generally used to provide heat to buildings.

Thermal solar systems

Solar energy as a source of heat primarily for heating and hot water is actively used using suitable collector systems. These systems change the solar radiation into heat and conduct it to the place of use via a

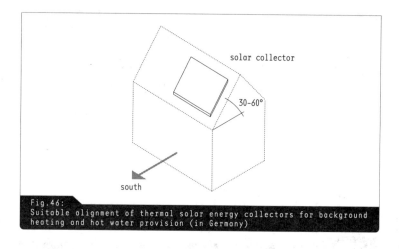

Fig. 46:
Suitable alignment of thermal solar energy collectors for background heating and hot water provision (in Germany)

heat-transporting medium (water with an antifreeze agent). There are two principal types of system, one using plate collectors, the other vacuum tube collectors. The latter are technically more complicated in construction, and therefore more expensive, but also more efficient.

If the solar energy system provides hot water only, the solar heat energy can be taken directly to the hot water storage tank. Any remaining energy demand for hot water provision is then normally covered by incorporating a more or less conventional boiler. Thermal solar energy systems can also be used as background heating for well-insulated houses.
> Fig. 45

Solar collectors are mainly positioned on roofs. The optimum alignment depends on the local course of the sun and the seasons of use. > Fig. 46 Small departures in orientation and inclination are tolerable and reduce the yield by only a small amount.

\\ Important:
Thermal solar collectors should not be confused with solar cells! Solar cells are also mainly installed on roofs but only generate electricity. This is either stored in batteries (island solution) and used directly on the premises or, if possible, fed into the public grid and payment is received.

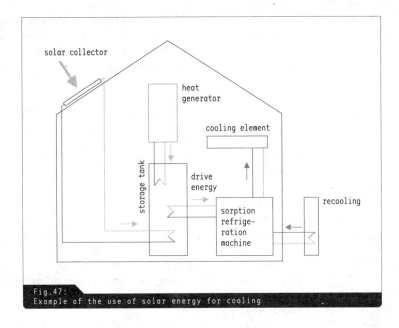

Fig. 47:
Example of the use of solar energy for cooling

Sorption refrigeration machines

Sorption refrigeration machines work using a similar principle to heat pumps, by compressing and expanding a medium. Instead of electricity or gas providing the auxiliary energy, a source of heat drives the system. The chemical process of sorption generates cold instead of heat.

Heat as the drive energy makes the use of sorption refrigeration machines advantageous where heat is available at no cost as a process waste (production plants, power-heat coupling plants etc.), or from the environment (thermal solar collectors, hot springs etc.), at a temperature of 80–160 °C. Combining a combined heat and power plant (CHP) with a sorption refrigeration machine for power-heat-cold coupling (PHCC) is perfect with respect to economy of operation, as one system covers the heat and cold energy demand, and can operate throughout the year.

**Local and district heat**

District heat is generated in central heating or heat-power plants as well as in decentralized combined heat and power plants. The latter solution is called local heat. Local and district heat can be used in remote buildings through an appropriate system of pipes and transfer stations. The generation of heat using the principle of power-heat coupling (CHP) is a low maintenance and environmentally friendly method of providing heat.

> Chapter Design principles, Covering the demands, Environmental impacts

**Heat transfer stations**

No heat generating equipment is necessary in the building to use local or district heat, nor any components such as heating equipment rooms, waste gas plants or fuel storage facilities. The heat is normally provided to the building in the form of hot water or steam through insulated pipes, and is transferred by a heat transfer station (heat exchanger) into the building's heating or hot water system.

**Process waste heat**

In some cases, superfluous heat from energy-intensive industrial processes (steelmaking, chemical industry etc.) can be fed into buildings through local and district heating networks. As this energy would otherwise be released into the environment, this option also improves the energetic efficiency of the industrial processes.

### Electricity

In principle, electricity can also be used to generate heat and cold. However, this option should be avoided, especially if the electricity comes from fossil fuels, because its generation already involves heat, and losses take place every time energy is converted (e.g. from coal to heat to electricity to heat). This can easily be seen from the primary energy factor for electricity for each location. › Chapter Design principles, Covering the demands, Environmental impacts

**Electrical heating**

Electrical heating systems are used only in exceptional circumstances. One example is bathrooms in existing buildings. Here, the heat demand considered over the year is small and a connection to a central heating system may be technically impossible or not worthwhile on economic grounds. The same applies to the electrical water heating in electrical boilers or immersion heaters.

Electricity is often used as auxiliary energy to drive heat pumps for heat generation. The use of electricity should be kept as low as possible to avoid high costs and a poor primary energy balance for the whole system.

**Compression refrigeration machines**

The most common cold generators are compression refrigeration machines. They work on the same principle as a refrigerator and generate cold from electricity, which then can be released to the building through a distribution system. With the use of appropriate energy, this can be a practical means of achieving any desired temperature level.

The waste heat from refrigeration machines can be used if there is a simultaneous need in the building for heat and cold (e.g. for room cooling and hot water provision). This waste heat is often released into the outdoor air so that, in combination with the (normally electrical) energy demand for the compressor, only a small amount of electricity is used.

There is an increased dependency on electricity as an energy source, which adversely affects the primary energy balance. Hence, the use of compression refrigeration machines should be avoided as far as possible.

## HEAT AND COLD STORAGE

By storing heat and cold it is possible to decouple use-dependent energy take-up from energy generation. This is necessary, in particular, in order to use solar energy for heating purposes, as the energy received from the sun varies according to the weather and often does not coincide with the times of day energy is required. > Fig. 48 Solar heat can be stored using the thermal storage capacity of the building itself, > Chapter Design principles, Covering the demands, Passive measures but this provides no possibility for users to regulate its use. Solar systems for hot water provision and background heating generally have a hot water storage tank, which may contain enough hot water for several days, but will take up space within the building.

Seasonal storage can be used to store solar heat over a longer period of time. This usually involves water tanks, which can be placed inside or outside the building. This allows heat to be stored in summer and made available in winter.

In principle it is also possible to store cold, e.g. using "ice tanks", which allows the whole system to be optimized in a similar way to heat storage. It is particularly the case with solar energy that the peak solar yield coincides with the greatest cooling demand. > Fig. 48

## HEAT AND COLD DISTRIBUTION
### Centralized systems

Distribution is the link between the generation of heat or cold and the system for transferring the heat or cold into the room, i.e. heat or cold is mostly generated centrally and carried to the transfer systems in the rooms by water or air as a transport medium. Water-operated systems are generally more viable with respect to energy content and the required

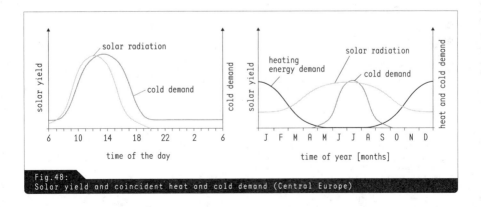

Fig. 48:
Solar yield and coincident heat and cold demand (Central Europe)

delivery energy, and should be preferred to air-operated systems in most situations. ❯ Chapter Design principles, Covering the demands The most common form of heating system in buildings is hot water-operated central heating. However, there are also design and economic factors in favor of using air-operated systems, which are discussed below.

The ductways should be kept as short as possible and be well insulated to minimize energy losses. The insulation also prevents water from condensing during the transport of cold media.

### Decentralized systems

Individual heating units generate heat energy directly in the room, where they also release it. There are various types of individual heating units, including stand-alone stoves, gas fires and electric fires. These are special forms of heating systems and are therefore not considered further in this publication.

Decentralized mechanical ventilation systems draw in outdoor air from directly in front of the facade, which dispenses with the need for ventilation equipment rooms and large duct networks. On the other hand, it costs more to maintain a large number of small devices, although this is mitigated by the units being increasingly modularized and simpler to maintain. They are normally also able to heat or cool the air required for ventilation. ❯ Chapter Tempering systems, Heat and cold transfer

### Regulation

In addition to insulated pipework, valves and pumps, there must also be a suitable means of regulating the distribution of heat and cold. These controls continuously match the output of heat and cold to the ever-changing demand caused by weather conditions (outdoor temperature,

\\ Example:
Buildings with a very high standard of thermal insulation, such as passive houses, have no need for a conventional heating system (with heating boiler, distribution and heat transfer by radiators). In spite of the physical disadvantages of air as a heat transfer medium, the required heat can be provided by a ventilation system, along with the minimum number of air changes necessary for hygiene.

\\ Note:
Radiator panels, which are used primarily for heating large halls, are one exception. They need to be operated at high feed temperatures.

wind, solar radiation), internal heat sources, and changes in room use. It must also be possible for the controls to act automatically in response to indoor and outdoor temperatures and be programmed with respect to time. In residential buildings the most common controls include thermostatic radiator valves, outdoor and room thermostats with suitable sensors, and automatic controls to reduce output at night at set times and in response to temperature changes. For larger buildings or systems, it may be worthwhile to install a building management system (BMS), which controls the heating or cooling in different rooms by means of sensors and a central computer.

## HEAT AND COLD TRANSFER

A suitable heat and cold transfer system uses water or air to carry the heat or cold from the generator to the tempered rooms where it is released.

Heat and cold transfer systems can be differentiated by the method of transfer, i.e. the proportions of radiated or convected energy, the feed temperature $T_f$, the specified capacity and controllability. > Chapter Design principles, Covering the demands

### Heat transfer systems

Heat may be transferred in a room by radiator, surface heating, or a ventilation system (air heating system). The construction and arrangement of the heating elements determine the temperature profile in the room, and have considerable influence on comfort. The transfer of heat in the room should be constant over time, and uniform in the horizontal and vertical directions to provide the ideal temperature profile. > Tab. 10, Fig. 49

*Heating radiators*

Common heat transfer systems in hot water-operated heating systems include radiators, convectors and surface heating. They are universal in use and can be easily controlled. Radiators and the like should be positioned on external walls and close to glazing to counteract the formation of draughts. > Fig. 50 One particular disadvantage of convectors is the high feed temperature required, which makes their use in combination with solar thermal systems, heat pumps or condensing boilers less viable.

*Surface heating*

Surface heating systems (floors, walls and ceiling heating), in which concealed heating pipes are buried in screeds, plaster or special panels, transfer their heat mainly by radiation into the room and create a comfortable temperature profile. > Fig. 49 The large heated area permits much lower feed temperatures than other heating bodies.

Surface heating is a suitable system for heat transfer in houses with a particularly low heating energy demand and in combination with low

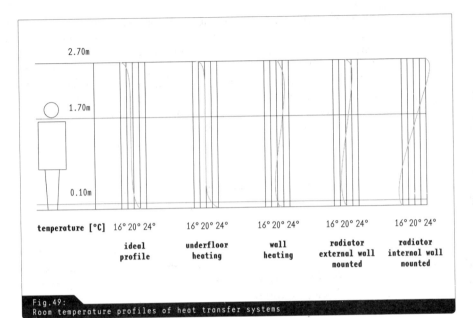

Fig.49:
Room temperature profiles of heat transfer systems

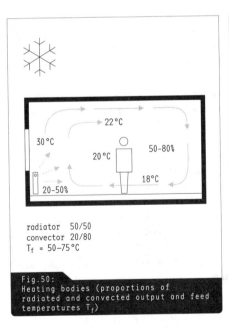

radiator 50/50
convector 20/80
$T_f = 50-75\,°C$

Fig.50:
Heating bodies (proportions of radiated and convected output and feed temperatures $T_f$)

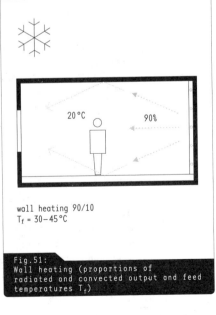

wall heating 90/10
$T_f = 30-45\,°C$

Fig.51:
Wall heating (proportions of radiated and convected output and feed temperatures $T_f$)

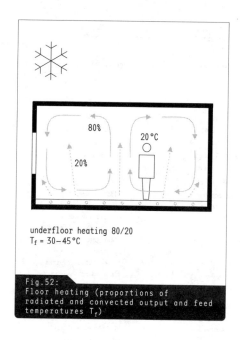

Fig.52:
Floor heating (proportions of radiated and convected output and feed temperatures $T_f$)

temperature systems, such as condensing boilers, heat pumps and coupled thermal solar energy systems. › Figs 51 and 52

Air heating

In principle, the heat demand of a building can also be covered by an existing ventilation system. The air can be provided either centrally or locally, mainly through electrically driven heating registers. This is, however, only worthwhile if a ventilation system is already required for other reasons and the building's overall heat energy demand is low (e.g. as a result of very good thermal insulation or high internal heat loads). Table 10 (see next page) gives a brief overview of the main heat transfer systems.

### Cold transfer systems

Transfer systems for cold energy, like those for heat energy, can be classified as convection or radiation systems according to their methods of transfer. The advantages and disadvantages arise from the effects on the operative room temperatures, as described in the chapter on Design principles, Comfort requirements. The risk of radiation asymmetry discussed there should be taken into account. In addition, controllability and the required feed temperatures are determining factors in the choice of a suitable cooling system. › Tab. 11

**Tab.10:**
Characteristics of heat transfer systems

| Heat transfer system | Advantages | Disadvantages | Proportion of output radiated/convected | $T_f$ [°C] |
|---|---|---|---|---|
| Radiators | Inexpensive, quick-acting, good controllability | Space required, appearance | 50/50 | 50–75 (90) |
| Convectors | Space-saving, quick-acting, good controllability | Cleaning, create dust | 20/80 | 50–75 (90) |
| Floor heating | Pleasant temperature profile, concealed | Cold air drop may occur, varicose veins may develop due to relatively high near floor temperatures, sluggish controls, not suitable for some floor coverings | 80/20 | 30–45 |
| Wall and ceiling heating | Pleasant temperature profile, concealed, heating and cooling possible | Sluggish controls, no furniture or fittings may be placed against heated walls, adequate ceiling height necessary for ceiling heating, people must keep a distance from the heated surfaces | 90/10 | 30–45 |
| Air heating | Combination of ventilation and heating, quick-acting controls | Draughts possible, dust created over 49 °C | 0/100 | 30–49 (70) |

As with the transfer of heat energy, cooling systems that exploit the mass of large building components are sluggish to respond to controls. Air-operated systems and systems that do not involve large building components respond quickly to controls and react more rapidly to changes in conditions. Large surface area heat transfer devices tend to have moderate (in the case of cooling: higher) feed temperatures than smaller devices, and are therefore better suited for use in combination with renewable energy sources.

A special point that must be kept in mind with cooling systems is the danger of falling below the dew point: depending on the temperature

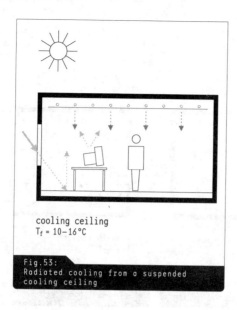

Fig. 53:
Radiated cooling from a suspended cooling ceiling

cooling ceiling
$T_f = 10-16\,°C$

and humidity of the room air and the surface temperatures of the cooling element surfaces, the moisture contained in the air may condense on the cold surfaces because their temperatures fall below the dew point. The condensate formed there must at least be collected; an even better solution is to drain it away. If this is not possible, condensation on cooled surfaces can be dealt with by temporarily increasing the surface temperature of the cooling body, which reduces its cooling capacity.

Cooling ceilings

Cooling ceilings can be found in many office and administration buildings. These transfer systems are suspended from, and may cover, large areas of the ceiling. › Fig. 53

Cooling coils filled with a flowing cold medium are suspended from the ceiling. About half of the cold transfer takes place by radiation, and the system responds well to controls. Feed temperatures are 10–16 °C and it is practically impossible to drain condensate away. To prevent condensation from forming, either high levels of humidity should be avoided or the feed temperature should be raised for a short period. In combination with natural ventilation in particular, these measures can mean the full cooling capacity is not available on warm, humid days.

Recirculated air cooling

A solution in widespread use is room cooling by air that has been extracted from the room itself, cooled, and then reintroduced into the room (recirculated air operation). These systems are normally decentralized and respond quickly to controls, which may be separate for each room. They

recirculated air cooling
$T_f = 6-10\ °C$

**Fig.54:**
Convective cooling by recirculated air units

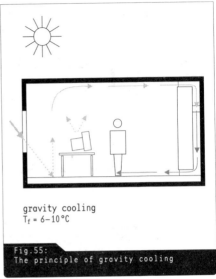

gravity cooling
$T_f = 6-10\ °C$

**Fig.55:**
The principle of gravity cooling

can also be controlled by the user. Air can be cooled by a cold register directly at the air outlet of a device suspended from the ceiling. The effect is purely convective. Feed temperatures of 6–10 °C are generally required.
› Fig. 54

Gravity cooling

Fully recirculated air cooling can also be performed without the need for a fan, using the principle of gravity cooling (or downdraught cooling). The movement of air is due to the fact that cold air is heavier than warm air. The room air is cooled by a cold convector near the ceiling (where the highest room air temperatures are found) and gradually drops down a shaft, accelerating the convection effect, before emerging at the bottom and spreading out across the room as a "pool of cold air." › Fig. 55

Gravity cooling offers the advantage of being completely silent. The cooling capacity in the room varies with temperature difference at the cooler and operates automatically to a certain extent. It can be installed inconspicuously behind a wall lining or a curtain without detrimental effect on its performance, as long as the top and bottom air entry and exit points are there. The problem of condensation can be solved as described in the systems above. A brief overview of the main cold transfer systems can be found in Table 11.

**Tab.11:**
Characteristics of cold transfer systems

| Cold transfer system | Advantages | Disadvantages | Proportion by radiation/convection | $T_f$ [°C] |
|---|---|---|---|---|
| Cooled ceilings | Moderate feed temperatures; rapid controllability; radiation effect; individual room control possible | Draining away condensation water is practically impossible | 50/50 | 10-16 |
| Recirculated air cooling | Rapid response to controls; individual room control possible | Purely convective effect; draining away condensation water is practically impossible; fan noise possible | 0/100 | 6-10 |
| Gravity cooling | Rapid response to controls; individual room control possible; completely silent; inconspicuous installation possible | Purely convective effect; draining away condensation water is difficult | 0/100 | 6-10 |

## Hybrid systems

A number of transfer systems are suitable for supplying heat and cold to a room. › Tab. 12

*Air conditioning equipment*
These systems also include air conditioning equipment that provides ventilation as well as heating and cooling a room from a central plant. › Fig. 56 The advantage of an air conditioning system is that it can humidify or dehumidify the room air, which means that practically any desired indoor air conditions can be achieved. › Chapter Ventilation systems, Mechanical ventilation

One fundamental disadvantage is that air is a poor heat transport medium, because of its low thermal storage capacity. › Chapter Design principles, Covering the demands Increasing the air volume flow to cover a demand for heat or cold energy therefore leads to increased energy use for delivering the air, which could be avoided in a water-operated system. Air conditioning systems often have no facility for allowing individual room control or user influence. This fact and the pronounced effect of air temperature, velocity and degree of turbulence on thermal comfort can result in user complaints and dissatisfaction.

*Thermally activated components*
Increased comfort can usually be created with systems that work by radiation, › Chapter Design principles, Comfort requirements for example with

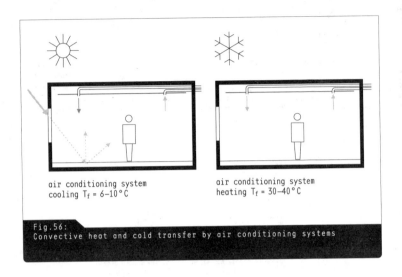

Fig. 56:
Convective heat and cold transfer by air conditioning systems

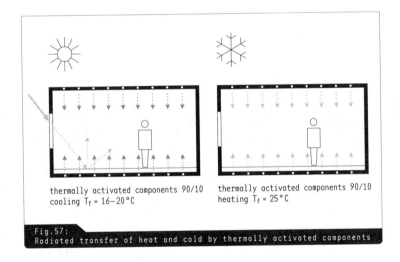

Fig. 57:
Radiated transfer of heat and cold by thermally activated components

thermally activated components. In these systems, pipework coils carrying a heat or cold medium are cast into the core of a concrete slab. › Fig. 57

The thermal inertia of the concrete mass makes controllability very low. Thermally activated components are therefore mainly used to cover constant loads or to meet a basic level of demand. › Chapter Design principles, Covering the demands

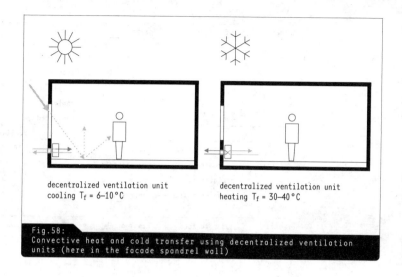

Fig. 58:
Convective heat and cold transfer using decentralized ventilation units (here in the facade spandrel wall)

For heating, the feed temperature is 25 °C. The performance of the activated component depends mainly on the temperature difference between the ceiling surface and the room air. This produces a self-regulating effect as the cooling output increases with rising room air temperature.

The problem of the dew point is similar to cooling ceilings, but the higher feed temperatures of 16–20 °C permit higher humidity without condensation taking place. The moderate feed temperatures for both heating and cooling modes mean that renewable energy sources can be used.

The coolant medium and the thermal mass of the concrete are integrated effectively, resulting in efficient de-energization of the component, for example overnight. Like night ventilation, this reduces › Chapter Design principles, Covering the demands the cooling output demand during the day and achieves a more uniform load profile. › Fig. 14, p. 28

The concrete surface must be thermally accessible for activated components to work properly. Thermally activated ceiling systems are incompatible with suspended ceilings. Impact sound insulation also reduces the effectiveness of these systems, as cold transfer on the top surface is limited.

Decentralized ventilation units

Decentralized ventilation units are built into the facade (e.g. in the spandrel wall), or in the facade area (e.g. in the double floor), and have a direct connection to the outdoor air. › Fig. 58

Controllability is very good, but air conditioning units require an increased flow of air to cover higher cooling loads, which means their use as the only means of heating or cooling is not always a workable option.

Tab.12:
Characteristics of hybrid systems for heat and cold transfer

| Hybrid system | Advantages | Disadvantages | Proportion of output radiated/convected | $T_f$ [°C] Heating | $T_f$ [°C] Cooling |
|---|---|---|---|---|---|
| Air conditioning systems | Rapid response to controls; draining away condensate simple | Purely convective effect; individual room control options poor | 0 / 100 | 30-40 | 6-10 |
| Thermally activated components | Moderate feed temperatures; radiation effect; partial self-regulation | Limited capacity; draining away condensation water practically impossible | 90 / 10 | 25 | 16-20 |
| Decentralized ventilation units | Rapid response to controls; individual room control easily possible | Draining away condensate difficult; increased maintenance cost | 0 / 100 | 30-40 | 6-10 |

If decentralized air conditioning units are used to cover higher cooling loads, then some means of draining away the condensate is necessary.

Decentralized systems can work well in combination with activated components, as the systems' advantages complement one another and their disadvantages cancel out.

As with centralized air conditioning systems, decentralized units normally require a feed temperature of 6–10 °C for cooling and 30–40 °C for heating. Table 12 shows the most common hybrid systems for heat and cold transfer. › Appendix, pp 84 and 85

### CHOOSING THE RIGHT SYSTEM

Selecting the most suitable tempering system to fit all demands often presents the architect with a difficult task. The various energy sources each require suitable heat and cold generators, but not every generator can be combined with every transfer system. It is often this complex interaction

**Tab.13:**
**Criteria in the selection of a tempering system**

| | |
|---|---|
| Technical criteria | Capacity (covering of demand) |
| | Suitable system temperatures of the components |
| | Suitability for renewable energy |
| | Availability of the energy source |
| | Possibility of heat recovery |
| | Controllability |
| Environmental effects | Primary energy demand |
| | $CO_2$ emission |
| User acceptance | Comfort requirements |
| | User influence |
| Economy | Initial costs |
| | Operating costs |

of the individual components and the number of more or less viable possible combinations that renders working closely with a specialist design engineer quite indispensable.

There is a series of further criteria relevant to the selection, which go beyond purely technical considerations. These are, primarily, energy efficiency, user acceptance and economic viability, all of which mean that the architect must proceed very carefully with the design.

Table 13 gives an overview of the most important criteria to consider in the choice of a suitable tempering system.

# COMBINATION OF VENTILATION AND TEMPERING

## THE RANGE OF POSSIBLE SOLUTIONS

The proposed ventilation and tempering systems must be positively combinable with one another so that overall concepts for indoor air conditioning systems can be developed that ensure the desired room temperature and required room ventilation.

Low-tech and high-tech

Depending on requirements, there are numerous possible combinations that vary in their degree of technical sophistication. Possible concepts for room conditioning range from the technologically simplest (low-tech) variants with window ventilation and radiators, through to the most complex (high-tech) systems with full air conditioning. › Fig. 59

Even though these systems represent the two extremes and therefore the start and end points of the range of possible solutions, nevertheless, depending on the requirements and the criteria applied, they may still be the most suitable concept for room conditioning, providing a performance to match the demand.

The number of possible combinations of system components precludes any universally applicable arrangements in the sense of a "patented solution" for room conditioning. A more fruitful approach is for the architect or specialist design engineer to evaluate the possible concepts based on the project criteria in order to arrive at a suitable concept.

## SELECTION CRITERIA

The combination of components for ventilation and heat and cold transfer can be evaluated according to technical criteria (e.g. renewable energies and heat recovery) and criteria relevant to user acceptance (e.g. comfort and user influence).

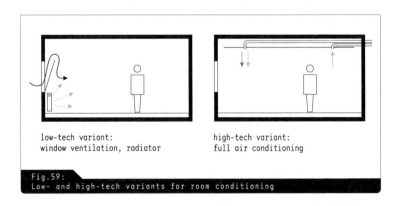

low-tech variant:
window ventilation, radiator

high-tech variant:
full air conditioning

Fig.59:
Low- and high-tech variants for room conditioning

# IN CONCLUSION

Buildings generally require individual solutions for room conditioning. In contrast to other industries, each component has to be checked against the specifically applicable framework conditions and requirements in detail and a customized solution devised. For this reason, achieving a successful concept usually involves a comparative examination of alternatives. Even having reached a satisfactory solution should not stop other possible combinations from being examined and their advantages and disadvantages evaluated.

Experienced architects and engineers are able to recognize advantages and disadvantages more quickly and therefore recommend particular applications. However, it is not possible to master the complexity of this subject without more accurate analysis.

This book therefore seeks to show that the detailed explanations of the individual system components and the examples of combinations are important as an overview. They provide an introduction to the subject of room conditioning in order to be able, later on, to create a fully designed room conditioning concept through calculations and technical drawings. However, it is only through understanding the dependencies, and being aware of the interaction of the specific project parameters with the technical possibilities, that an optimum solution can be found for the project.

# APPENDIX

## EXAMPLES OF CONCEPTS

### Window ventilation, radiators

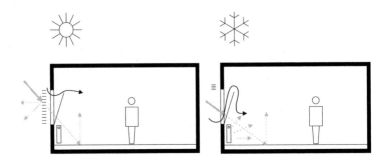

Window ventilation with radiators + solar screening

| Suitability | Advantages | Disadvantages |
|---|---|---|
| _ Residential<br>_ Office | _ Efficient energy transport by water<br>_ User influence and individual room control easily possible<br>_ Decoupling of ventilation and tempering | _ Unconditioned supply air<br>_ No defined air change rate possible<br>_ No heat recovery possible<br>_ No cooling, humidification or dehumidification possible<br>_ Uncomfortable and high heat losses in winter<br>_ Noise and dust emissions possible |

## Window ventilation, underfloor heating

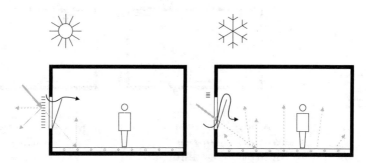

Window ventilation with underfloor heating + solar screening

| Suitability | Advantages | Disadvantages |
|---|---|---|
| _ Residential<br>_ Office | _ Efficient energy transport by water<br>_ Suitable for the use of renewable energy<br>_ User influence and individual room control easily possible<br>_ Comfortable heat transfer<br>_ Decoupling of ventilation and tempering | _ Unconditioned supply air<br>_ No defined air change rate possible<br>_ No heat recovery possible<br>_ No cooling, humidification or dehumidification possible<br>_ High heat losses in winter<br>_ Noise and dust emissions possible<br>_ Very sluggish response to controls |

## Air heating

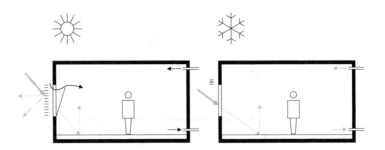

Air heating with optional window ventilation in summer + solar screening

| Suitability | Advantages | Disadvantages |
|---|---|---|
| _ Residential | _ No additional heating surfaces required | _ Inefficient energy transport by air |
| | _ Defined air change rate possible | _ Only efficient with low heat demand |
| | _ Heat recovery possible | _ Only convective heat transfer |
| | _ Noise and dust emissions can be avoided | _ User influence and individual room control difficult to incorporate |
| | | _ Coupling of ventilation and tempering |

## Window ventilation, convectors, cooling ceilings

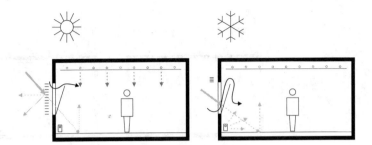

Window ventilation, convectors, cooling ceilings + solar screening

| Suitability | Advantages | Disadvantages |
|---|---|---|
| _ Office<br>_ Conference room | _ High comfort in summer<br>_ Efficient energy transport by water<br>_ User influence and individual room control easily possible<br>_ Decoupling of ventilation and tempering | _ Unconditioned supply air<br>_ No defined air change rate possible<br>_ No heat recovery possible<br>_ No humidification and dehumidification possible<br>_ Condensate drainage difficult to incorporate<br>_ Discomfort and high heat losses in winter<br>_ Noise and dust emissions possible |

## Window ventilation, convectors, recirculated air units

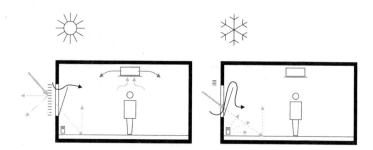

Window ventilation, convectors, recirculated air units + solar screening

| Suitability | Advantages | Disadvantages |
|---|---|---|
| _ Office | _ User influence and individual room control easily possible<br>_ Decoupling of ventilation and tempering | _ Unconditioned supply air<br>_ No heat recovery possible<br>_ Only convective heat and cold transfer<br>_ Possible discomfort and high heat losses in winter<br>_ Noise and dust emissions possible |

## Decentralized ventilation systems, thermally activated components

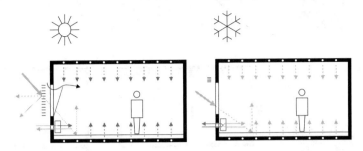

Decentralized ventilation systems, thermally activated components with optional window ventilation in summer + solar screening

| Suitability | Advantages | Disadvantages |
|---|---|---|
| _ Office<br>_ Conference room | _ Efficient energy transport by water (basic load cover)<br>_ Heat and cold transfer partially by radiation<br>_ Preconditioned supply air<br>_ Defined air change rate possible<br>_ Heat recovery possible<br>_ User influence and individual room control possible<br>_ Decoupling of ventilation and tempering<br>_ Noise and dust emissions can be avoided | _ Higher installation costs<br>_ Condensate drainage difficult to incorporate<br>_ No humidification and dehumidification possible<br>_ Higher maintenance costs |

## Air conditioning systems

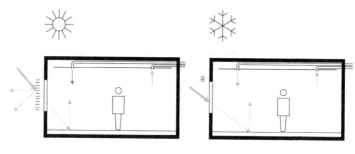

Air conditioning system + solar screening

| Suitability | Advantages | Disadvantages |
|---|---|---|
| – Residential<br>– Office<br>– Conference room | – Preconditioned supply air<br>– Defined air change rate possible<br>– Heating, cooling, humidification and dehumidification possible<br>– Heat recovery possible<br>– Condensate drainage easy to incorporate<br>– Noise and dust emissions can be avoided | – Inefficient energy transport by air<br>– Only convective heat and cold transfer<br>– User influence and individual room control difficult to incorporate<br>– Coupling of ventilation and tempering |

The following table gives an overview of possible combinations of the ventilation and tempering systems presented in this book. This table is neither generally applicable nor exhaustive, but should be used as an example of how the various concepts can be compared and a suitable solution found. The concepts selected as examples are also shown in detail in the appendix to this book.

| room conditioning system | ventilation type | | tempering type | | | | | | | | characteristics | | | | | typical use | | |
|---|---|---|---|---|---|---|---|---|---|---|---|---|---|---|---|---|---|---|
| | natural | mechanical | heating bodies | surface heating | air heating | cooling ceiling | recirculated air cooling | air conditioning system | thermally activated components | decentralized ventilation unit | suitable for renewable energy | heat recovery | controllability | comfort | user influence | residential | office | conference room |
| | X | | X | | | | | | | | o | – | + | o | + | X | X | |
| | X | | | X | | | | | | | + | – | – | + | + | X | (X) | |
| | (X) | X | | | X | | | | | | – | + | + | o | – | X | | |
| | X | | X | | | X | | | | | o | – | + | + | + | | X | X |
| | X | | X | | | | X | | | | o | – | + | o | + | | X | X |
| | (X) | X | | | | | | | X | X | o | + | o | + | + | | X | X |
| | | X | | | | | | X | | | – | + | + | – | – | (X) | X | X |

# STANDARDS

## Standards and guidelines referred to in the text:

| | |
|---|---|
| DIN 1946 | "Ventilation and air conditioning", Part 2 "Technical health requirements (VDI ventilation rules)", 1994-01 (withdrawn) |
| DIN EN 12831 | "Heating systems in buildings - Method for calculation of the design heat load", Supp.1 "National Annex", 2006-09 |
| DIN EN 15251 | "Indoor environmental input parameters for design and assessment of energy performance of buildings addressing indoor air quality, thermal environment, lighting and acoustics", 2007-08 |
| DIN EN ISO 7730 | "Ergonomics of the thermal environment - Analytical determination and interpretation of thermal comfort using calculation of the PMV and PPD indices and local thermal comfort criteria (ISO 7730:2005)", 2006-05 with Amendment 2007-06 |
| DIN V 18599 | "Energy efficiency of buildings - Calculation of the net, final and primary energy demand for heating, cooling, ventilation, domestic hot water and lighting", Part 1 "General balancing procedures, terms and definitions, zoning and evaluation of energy sources", 2007-02 |
| VDI 2078 | Technical regulation "Cooling load calculation of air-conditioned rooms (VDI cooling load regulations)", 1996-07 |
| VDI 3803 | Technical regulation "Air-conditioning systems - Structural and technical principles", 2002-10 |

## LITERATURE

Hazim B. Awbi: *Ventilation of Buildings*, E & FN Spon, London 1991
Sophia and Stefan Behling: *Solar Power*, Prestel, Munich 1996
Klaus Daniels: *Advanced Building Systems*, Birkhäuser Verlag, Basel 2003
Klaus Daniels: *Low-Tech Light-Tech High-Tech*, Birkhäuser Verlag, Basel 1998
Baruch Givoni: *Climate Considerations in Building and Urban Design*, John Wiley & Sons, New York 1998
Baruch Givoni: *Passive and Low Energy Cooling of Buildings*, John Wiley & Sons, New York 1994
Gerhard Hausladen, Michael de Saldanha, Petra Liedl, Christina Sager: *Climate Design*, Birkhäuser Verlag, Basel 2005
Norbert Lechner: *Heating, Cooling, Lighting: Design Methods for Architects*, John Wiley & Sons, New York 1991
Christian Schittich (ed.): *Solar Architecture*, Birkhäuser Verlag, Basel 2003
Steven V. Szokolay: *Environmental Science Handbook for Architects and Builders*, The construction press/Lancaster, London 1980
Steven V. Szokolay: *Introduction to Architectural Science: The Basis of Sustainable Design*, Elsevier Architectural Press, Oxford 2004

## THE AUTHORS

Oliver Klein, Dipl.-Ing., is an architect and energy consultant. He has taught and researched as a post-graduate assistant at the Faculty of Climate-compatible Architecture at the Dortmund University of Technology since 2005.

Jörg Schlenger, Dipl.-Ing., is a construction engineer and consulting engineer for energy and air conditioning concepts, specializing in thermal and energetic building simulation. He has been a post-graduate assistant at the Faculty of Climate-compatible Architecture at the Dortmund University of Technology since 2004.

## 导言

**人体的温度**

虽然很多物种能够调节他们的体温来适应其周围的环境,但人类却需要一个几乎恒定的身体温度($37 \pm 0.8$℃)。当外界温度随着气候区域、一天中的时间和季节波动时,人体尝试利用自身的自动调节系统来维持这个温度,在这一过程中皮肤表面根据环境温度与身体活动水平释放出或多或少的热量。例如,如果人体温度升高,那么汗腺允许汗液暴露在皮肤表面并通过蒸发将人体释放出的热量带到环境中。如果人体温度降低,皮肤会紧缩以减少释放热量的面积,相应的皮肤上的毛发(汗毛)会立起来("起鸡皮疙瘩");同时,人体也会通过肌肉的颤动(发抖)创造额外的热量。

**气候的影响和补偿**

然而,这个温度调节系统也是有限的,因为人体皮肤仅能在一定程度上完成这项任务。服装作为附加的"保温",可视为人的"第二皮肤";而建筑可视为人的"第三层皮肤",它们为拓展这个温度调节系统的适用范围提供了解决方案。

在人类历史上,火的发明无疑是人类成功摆脱对气候与季节依赖性的最重要的一步。它不仅标志着人类进入了化石能源时代,即能量的转化依赖于持续的能源供应;从此,"第三层皮肤"就有了可以使用的人造热和光——因此,可将它视作室内环境调节的雏形。现如今,使用化石能源带来的环境破坏问题已经广为人知且无处不在。

**能源优化的室内环境调节**

术语"室内环境调节"可以理解为:为了使人们享受舒适的感觉,通过调温(加热或制冷)、照明和引入足够的新鲜空气(通风),创造一个不受任何外界影响的室内环境;利用合适的技术,这可导致最终统一且与建筑位置无关的系统架构。举一个极端的例子:这些建筑被密封在玻璃幕墙中,广泛地采用高新技术实现幕墙内部全面的空调控温,进而可发现——在世界上所有的气候区域,这些建筑都可以按照非常一致的设计来建造。除干扰用户的敏感性之外,这类系统的另一缺点是供暖、供冷和照明的能量需求非常高。世界范围内50%的能量被用于建筑这一事实证明,必须构想其他的方法来提供能源最优的室内环境调节。

一个建筑总是应该被设计成以最少的额外能量消耗来提供所需的舒适性。首先,在转向技术(主动式)措施之前,所有可用于室内环境调节的构造(被动式)措施都应结合地域条件予以综合开发利用(参见"设计原理"一章)。

为获得室内环境调节的能源优化整体概念，在被动式方法和主动式方法的有效结合中，保证所有技术组件之间可以相互兼容总是至关重要的。接下来的章节讲解室内环境调节的基本原则及其相互作用，并尝试使设计师们能够针对每一个建造项目开发出独特和均衡的室内环境调节系统。

P10 # 设计原理

P10

## 1.1 舒适性需求

### 1.1.1 热舒适

术语"舒适"描述了一种健康、幸福的感觉，它受多种因素影响。在建筑设备系统领域，舒适通常指热舒适。热舒适描述的是人体热平衡与周边气候条件达到均衡的一种状态，在这种状态下，用户感觉到周边的气候条件既不是太热，也不是太冷。

舒适性的重要性

热舒适并不奢侈，它是用来评价建筑能否实现其设计目标的一个非常重要的准则。在一个建筑中，空间的质量对用户集中精力并高效工作的能力，以及其健康状况（例如在办公室）具有多重影响。如果生产区的舒适度不够，则可能使早疲劳的问题增加，进而为工作安全埋下隐患。因此，提供可靠且与应用相符的室内环境是成功建筑的一个非常重要的质量特性。

### 1.1.2 影响因素

我们对于舒适的感知依赖于很多影响因素，如图1所示。

一栋建筑的设计者们通常只能够影响到其中的物理状态，关于物理状态的一些内容我们将在后续章节中详细描述。然而，用户的衣着与活动水平会明显地影响到他（或她）对室内环境舒适度的评价，这些因素和用户自身的调节与适应能力都属于"中间"因素这一组，并且同时受物理与生理这两种条件的影响。

在有些案例中，其他因素也可能在建筑的初步设计阶段起到重要作用，这些知识与有意识地专门为指定用户群创造一个建筑的概念，往往是设计过程中不可或缺的。例如，老年人感觉舒适的温度通常高于大众，因此，在为老人设计护理房间时，就必须考虑这一因素，适当地将室内温度水平提高到老年人感觉舒适的区间。

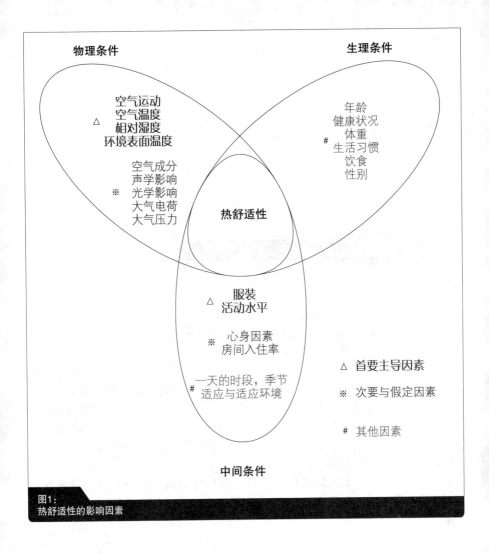

图1:
热舒适性的影响因素

### 1.1.3 物理状态

空气和辐射温度

人的身体也和其他物体一样，持续不断地通过辐射的方式与其周围表面交换热量。因此，除空气温度之外，最重要的物理状态参数是周围表面的平均温度。在辐射换热中，或多或少的热量将由一

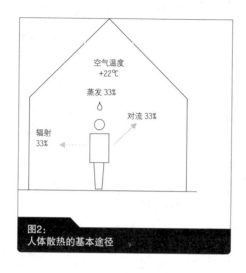

**图2:**
**人体散热的基本途径**

个或者其他方向散失或吸收，辐射换热量的大小与热量净传递方向取决于参与换热的两主体之间的距离与温差。因此，辐射换热过程会影响人体的热平衡。

空气温度与辐射温度之间的差别较小时，依然能够满足人体对热舒适度的要求。如果空气和辐射温度之间的差异或不同辐射温度之间的差别过大，就会导致不舒适。这也正是在室内温度很适宜的情况下，当你站在一个非常热或非常冷的辐射面（例如，一个隔热不善的建筑构件或窗户）附近时，会感觉到不舒适的原因（参见图 3）。

> 操作温度

由于人的身体不能够探测出绝对温度，而只能通过皮肤或多或少地感觉到强烈的热损失或得热，因此人体对温度的敏感

> 提示：
> 在少量身体活动、正常衣着与常规室内温度下，人体通过辐射、对流、蒸发散失的热量均为人体总散热量的三分之一（如图 2 所示）。

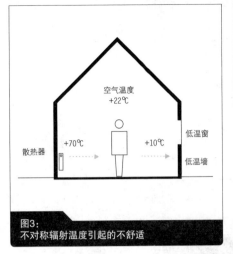

图3：
不对称辐射温度引起的不舒适

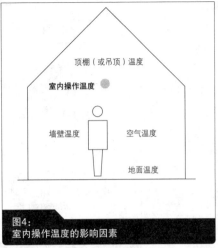

图4：
室内操作温度的影响因素

度依赖于人体与室内空气和周围表面的热交换量。空气温度和辐射温度二者对人体热感觉的综合影响以"操作温度"（或"感知温度"）来表示，操作温度已经成为评价热舒适性的一项权威设计指标。

室内不同位置处的室内操作温度是不同的，其量值取决于离房间表面的距离。为了便于设计，通过空气温度的均值与室内各表面辐射温度的均值来计算其量值，并采用该数值来计算室内的热状态。

湿度

因为人体通过蒸发散失部分热量，所以另一个影响我们舒适性感觉的因素是空气的相对湿度。受湿度影响，一个给定温度状态在

建议：

当人员活动以久坐不动为主时，例如在办公室或家中，顶棚太热或者是墙或窗户的表面太冷，都会很快的引起不舒适。房屋表面与室内空气之间的温差不应该超过3开尔文（K），在一定的范围内，高一些的室内空气温度可以弥补低一些的表面温度，反之亦然。

提示：

操作温度推荐值范围的变化依赖于应用，同时也存在于国际比较中。例如欧洲，根据欧洲标准EN15251，对于轻微、久坐的活动，推荐的操作温度范围为20℃~26℃；而其他很多国家都有他们各自的规范，且这些规范中操作温度的推荐值范围常常与欧洲标准不同。

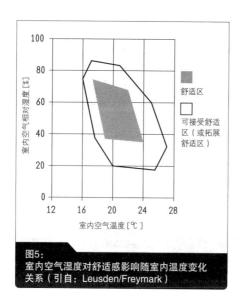

图5:
室内空气湿度对舒适感影响随室内温度变化关系（引自：Leusden/Freymark）

保持其他边界条件不变的情况下，可能被感知为热或冷（如图5所示）。

除了散湿负荷（室内的人和植物）外，室内相对湿度的另一个影响因素是室外空气的湿度（例如气候）。同时，相对湿度也随着室内空气温度的波动而变化。因此，在冬天加热引入室内的室外冷空气，一般会导致室内相对湿度变低。

> 提示：
> 对于大多数应用，常常推荐使用约为50%（±15%）这一相对湿度（例如，在家或办公室中轻微久坐的活动强度）。然而，必须考虑到，极少数室内环境调节系统（例如全空气空调系统）允许用户全权控制相对湿度。

**空气运动**　　空气运动能够独立于湿度，对热舒适性产生巨大的影响。高速运动的空气或涡流（空气流中高度的湍动）均可引起吹风感（或气流）。此时的舒适感依赖于空气的温度：空气温度越低，空气运动就会越快地被认定为不舒适；而温暖空气的运动则能够被容忍相当长的时间。空气温度越高，强涡流的数量越少是一个问题。

**空气成分**　　尽管地表水平面的空气成分通常是恒定的，但它还是能够影响总体的舒适感。这是因为，空气中的污染水平与依赖温度的水分含量是广泛变化的，并且都可能在特殊情况出现。

空气污染物可能源自建筑的外部与内部。室外的影响因素主要是建筑物的位置（道路交通等），和通风口的位置与尺寸的函数；室内的主要影响因素可以确定为建筑的排放物（建筑材料、建筑中的材料或使用建筑所释放出的气体），和建筑中的人的排放物（水蒸气、气味、呼出空气中的 $CO_2$ 等）。除去这些排放物与/或提供足够的空气交换，就应该能够使空气中各种物质含量遵守任何规定限额，与此同时，使不舒适——或损害健康——得以避免。

**空气中二氧化碳含量**　　室内空气中的 $CO_2$ 含量极其重要。$CO_2$ 随着室内人员的呼吸加到室内，当其含量仅仅达到千分之几之后，就会成为问题。与公众观点相反，导致空气是"坏的"或者"不新鲜"的并非是缺少氧气（$O_2$），而是 $CO_2$ 增多，进而产生了空气交换的需求。图6展示了一个教室中的 $CO_2$ 随着使用时间而增加。

**建议：**

在通过数量足够多且大的配风口通风的家或办公室中，对于轻微、久坐的活动，通常推荐采用的平均空气流速限制性条件为：小的湍流条件下（大概为5%）平均流速不超过 0.2m/s。高的流速会引起局部区域不舒适，例如排风口附近直接与出风方向相对的区域，这点也必须被充分考虑。还有更多明确的信息，可参考欧洲标准 EN 15251。

**提示：**

干空气由 78.1% 氮气（$N_2$）、20.93% 氧气（$O_2$）、0.03% 二氧化碳（$CO_2$）、0.94% 氩气以及其他惰性气体组成。此外，"正常的空气"中含有不同比例的水蒸气以及氮氧化物、二氧化硫、废气、灰尘、悬浮颗粒及多种微生物等形式的污染物。

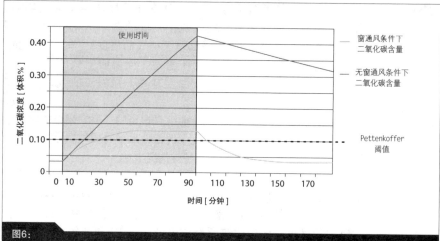

图6:
有无倾斜窗通风（换气次数为每小时3次）两种情况下教室（150m³，30人）内$CO_2$浓度在一段使用时间（90分钟）之中与之后的变化

其他污染物　　如果我们呼吸的空气被额外的污染物所污染，例如，来自吸烟、使用材料或工作区域制造过程所释放的污染物，那么对新鲜空气就有了额外的需求。这一要求由污染物的浓度所决定，例如，必须通过提供足够的换气次数来保证室内空气品质保持在允许的限度之内。

提示：

"Pettenkoffer 阈值"由慕尼黑医生与卫生学家 Max von Pettenkoffer 建立，其指出室内空气的 $CO_2$ 含量不应该超过 0.10%。在绝大多数欧洲国家中，均采用这一比较严格的要求作为评价室内空气品质的一个舒适性准则。因此必须采用适当的通风手段防止室内 $CO_2$ 含量超出这个阈值。

提示：

建筑设计中，仅在近期才考虑人的适应能力。在新的规范中，例如欧洲标准 EN15251，所推荐的舒适范围因此而根据同时发生的室外空气温度来界定。期待其他国家也能够将这一新知识吸收到他们的设计准则之中。

**人在从事不同强度水平的活动时释放热量值的范例**　　表1

| 活动 | 每人释放总热量的参考值（W） |
|---|---|
| 坐姿进行的固定的活动，例如阅读和书写 | 120 |
| 站姿或坐姿进行的非常轻的体力活动 | 150 |
| 轻体力活动 | 190 |
| 中度到高强度的体力活动 | 大于270 |

#### 1.1.4 中间条件

**用户活动**　　当人们从事体力活动时，人体需要消耗更多的能量，并需要向环境释放热量，参见表1。动作越少消耗的能量就越少，并且也不需要释放大量的热量。因此，用户的活动强度等级对他的舒适感有很大的影响。活动强度的等级越低，所需的舒适温度就必须越高。此外，人在从事轻体力活动时（例如久坐在桌旁），普遍的要比从事重体力活动（运动、体力工作等等）时，更能忍受"理想温度"的较小的偏离。

**服装**　　除了体力活动，服装影响人体释放热量的总量，进而影响人体的热平衡。穿的衣服越少，人体释放到空气以及周围表面的热量就越多。因此，在人光着脚或光着身体的区域（例如淋浴间或者桑拿间）往往需要提供较高的温度和温暖的表面（例如地板供暖），以避免不舒适。无论如何，由于衣服的保温效果，穿上更多的衣服总能够降低人们对周围环境温度波动的敏感度。

**适应能力**　　人体在短期、中期或者长期内适应特殊气候条件（例如，加热过程）的能力被描述为适应性或环境适应性。因此，这一适应能力可以意味着：最初感觉不舒适的环境在一段时间后可能被认为是舒适的或至少不是那么令人痛苦的。

### 1.2 确定需求

建筑的主要需求之一是根据指定的标准，在全年时间范围内保证一个适宜的、舒适的室内环境。当这些需求应用于所有不同的气候带时，所需的措施会因地带的不同而不同。

#### 1.2.1 通风需求

为保证一个建筑内的空气满足卫生质量要求，必须在规定的时间间隔内，通过引入新风置换需要引出的使用过的空气和污染物，

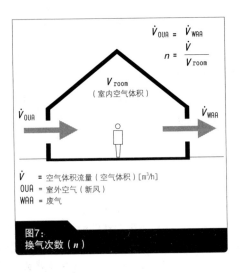

图7:
换气次数($n$)

以实现室内空气的更新。这一般通过在外墙上设置开启的窗来实现,但也经常通过机械通风系统来实现,参见"通风系统"一章。采用新风置换室内全部空气的频次,以及采用什么周期(例如一小时),可以通过换气次数($n$)来确定,参见图7。

> 需求的最小换气次数

为保证健康的室内空气,必须首先从屋子的使用功能角度来考虑室内空气所必需的换气次数。不同专业文献中给出的换气次数值往往不同,因此,对于典型的房间、人员密度与污染物负荷应该仅对其均值加以考虑。这对于初期设计是非常有用的。参见表2。

**提示:**

新鲜空气(新风)是指源于没有被污染的自然环境的空气。在通风工程中,它与室外空气不同,室外空气是从建筑外吸入的,而建筑外的空气可能是热的或者是被污染的。一般室外空气中会有一定浓度的 $CO_2$ 以及空气传播的污染物,但是其含量比室内空气中的量低。新风需求通常可通过引入"室外空气"满足。因此,后续本书将采用"室外空气"这一术语*。

**示例:**

例如一个面积为 20m² 的起居室,其净空高度为 2.50m,因此其体积为 50m³,起居室内有一人。根据表3,在一个小时内用室外清新空气替换室内空气的一半就足够了($n$=0.5/h)。表7(第40页)展示了一个通风装置(例如窗)需要开启多长时间。如果起居室中有两个人,那么所需的室外空气流量 $\dot{V}$ 为 50m³/h,因此,所需的换气次数为 1.0/h。也就是说,每小时室内空气都必须被室外空气完全替换一次。

*译者注:译稿采用国内普遍使用的"新风"来替代"室外空气"这一术语

**新风流量**

最好考虑预期的污染物负荷与具体的排放量,计算出单位时间内需要引入到室内的新风量,即新风流量 $\dot{V}$(通常用 $m^3/h$ 表示)。通过上述新风流量与房间体积,可以计算出换气次数,作为一个附加的设计参数使用。

在室内,如果空气没有严重受到建筑材料或特殊应用排放的污染物影响,那么房屋的使用人员成为主要的影响因素。

如前所述(参见"设计原理"一章,"舒适性需求"一节),一个人的新风量需求可以通过以 $CO_2$ 浓度未超过推荐上限值的空气来置换室内空气来满足。所需的新风流量取决于房屋的使用类型、预期的污染负荷以及室内人员的数量。在实践中,可以应用表中的数值,参见第 100 页表 3。

**不同空气流量**

应该采用满足卫生要求的最小换气次数(参见表 2)进行空气置换,尤其是在室外温度很低时,因为增加新风的供给量总会导致更高的通风热量损耗。另一方面,室内温度较高时也可以通过较高的换气次数来移除多余的热量。

### 1.2.2 调温需求

调温需求(热能量和冷能量的需求)总是由一个建筑或房间能量平衡的扰动引起。

推荐的换气次数 ($n$)　　　　表 2

| 房间类型 | $n$ 1/h |
|---|---|
| 起居室 | 0.6~0.7 |
| 卫生间 | 2~4 |
| 办公室 | 4~8 |
| 食堂 | 6~8 |
| 酒吧 | 4~12 |
| 电影院 | 4~8 |
| 报告厅 | 6~8 |
| 会议室 | 6~12 |
| 百货公司 | 4~6 |

基于 DIN EN 15251（Cat. Ⅱ）最小新风量　　　表3

| 房间类型 | $\dot{V}_{OUR\,min}$ m³/（h·人） |
|---|---|
| 起居室 | 25 |
| 电影院，音乐厅，博物馆，阅读室，体育馆，商用场所 | 20 |
| 独立办公室，礼堂，教室，小会议室，大会议室 | 30 |
| 酒吧 | 40 |
| 开敞式办公室（没有隔间的办公室） | 50 |

**一个房间的能量平衡**

　　室内温度应该保持在一个范围，一个对房间的使用来说可被认为是舒适的范围。建筑中的热量以热传导（室内外温差引起的通过建筑维护结构的热流）与通风（通过建筑维护结构中的不可控渗漏点与未控制的自然通风或机械通风进行的室内外空气交换）的方式被抽出（或散失到室外环境）。通过透明建筑构件进入的太阳辐射、设备发热和建筑中的人将热量带入建筑，参见图8。

　　尽管已针对地方气候对建筑进行设计（例如，一个高度隔热的建筑维护结构、遮阳等等），但为了达到目标室内温度，仍然有必要依据建筑的位置，时不时地以人工方式将热量引入或将多余的热量移出建筑或房间（通过建筑设备系统）。因为这一过程总是与额外的能量需求相关，所以其目标应该是通过采用与当地气候条件相匹配的智能建筑和建筑设备技术，来避免这一附加的调温能量需求或保持其尽可能小。参见"设计原理"一章，"满足需求"一节中的回避原则。

**供暖能量需求**

　　如果日照或内部得热是不够的，那么一个建筑或房间的热量损失必须因此而得到一个源自供热系统的对应量值的热量补偿（参见图9）。一栋建筑的年供暖能量需求，是在考虑各种相关的得热量情况下，计算出的建筑全年热量损失之和。为评估一栋建筑的热性能，需要计算出其使用空间每平方米的年供暖热量需求 kWh/（m²·a）。这一能量参数使建筑能够在能源方面按照国家能源标准进行描述与分类。

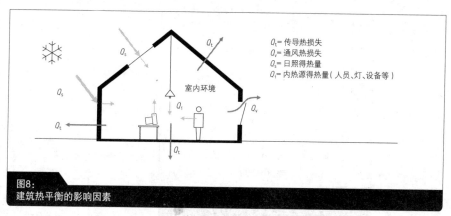

**图8:**
建筑热平衡的影响因素

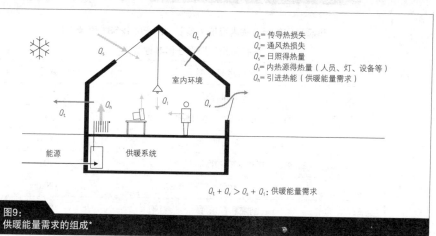

**图9:**
供暖能量需求的组成*

> **提示:**
> 年供暖能量需求并不包含必须供给建筑的能量的实际量值信息。为了计算出最终的能量需求 $Q_e$——它通常可以通过能量计量表监测（煤气或者电能表等等），连接到供暖系统的任何热水系统的影响，连同供暖系统效率，包括热分配相关损失、运行系统所需的额外能量（如泵所需电能等）等方面都必须加以考虑。为确定总体能量需求，包含其对一栋建筑所处环境的影响，必须考虑燃料的使用量，通过计算 $CO_2$ 排放量或总一次能源需求量（参见"设计原理"一章，"满足需求"一节，环境影响）。

\* 译者注：本图删除了 21 页 Fig.9 中的两处 $Q_t$。因为此两处 $Q_t$ 与室内的热负荷并不直接相关，故删除，以免产生误解

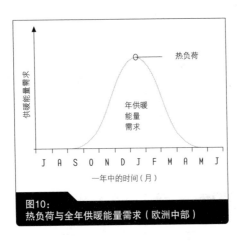

**图10:**
**热负荷与全年供暖能量需求（欧洲中部）**

供暖房间的标准室内温度（参照 DIN EN 12831 附录 1），
没有为顾客提供任何其他数值                                    表4

| 建筑或房间类型 | 操作温度 [℃] |
| --- | --- |
| 起居室和卧室 | +20.0 |
| 办公室、会议室、陈列室、主要楼梯井、售票厅 | +20.0 |
| 宾馆房间 | +20.0 |
| 接待大堂和商店 | +20.0 |
| 教室 | +20.0 |
| 剧院和音乐厅 | +20.0 |
| 厕所 | +20.0 |
| 洗浴和淋浴室、浴室、更衣室、检查室（总之，包含不着装的房间） | +24.0 |
| 供暖的辅助房间（走廊、楼梯井） | +15.0 |
| 不供暖的辅助房间（地窖、楼梯、储藏室） | +10.0 |

| | |
| --- | --- |
| 热负荷 | 为了设计供暖系统（热源与热分配系统），首先需要确定热负荷，从而确定所需的最大热量输出。这一数值表明了在一年中最冷的一天，为抵消热损耗以达到室内温度设计需求，必须供给建筑的热量（单位为 W 或 kW）。参见表4及图10。 |
| 制冷能量需求 | 当得热总量（进入建筑中太阳辐射以及内热源发热量的结果） |

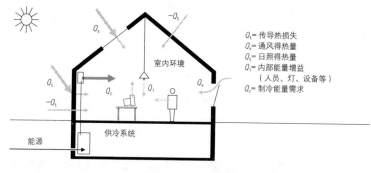

**图11:**
制冷能量需求的产生*

$Q_s + Q_i + Q_v > Q_t$：制冷能量需求

图中标注：
- $Q_t$ = 传导热损失
- $Q_v$ = 通风得热量
- $Q_s$ = 日照得热量
- $Q_i$ = 内部能量增益（人员、灯、设备等）
- $Q_c$ = 制冷能量需求

室内环境；供冷系统；能源

> 制冷能量需求计算

超过热量损耗总量（通风和传导损失的结果），且多余的热量不能被有效地存储在蓄热体中时，就产生了制冷能量需求。（参见"设计原理"一章，"满足需求"一节）。室内温度升高并用主动式方法（例如建筑设备系统）不能够防止室内温度超出最高设计温度限值，参见图11。

所包含的因素之间极其复杂的物理相互作用，使准确预测预期的制冷能量需求非常困难。除外部和内部负荷之外，设计人员还必须考虑建筑构造等蓄热体吸纳与释放能量的过程。为能够精确分析一个以小时为模拟步长的重要或困难问题的状况，通常必须借助动态模拟软件。

因此，在正常的设计情况下，通常不会做精准的制冷能量负荷估计，而是以基于静态（恒定）工况的制冷负荷计算取而代之。特别是这种计算方法允许估计最大冷负荷，这在供冷系统设计中非常有用。

> 提示：
> 当室外温度很高，除进入的太阳辐射与内热源（人体、设备、光照）引起的得热量之外，来自通风和保温不良的建筑构件的得热也对制冷能量需求有所贡献。

*译者注：本图相较 23 页 Fig.11 有部分改动。本图中"$Q_t$"有四项，其中上面两项（有人的房间）的 $Q_t$ 实为"传导得热量"；而在下面无人的一室，其对应 $Q_t$ 为"传导热损失"。由此可见，$Q_t$ 同时对应"得热"与"热损失"，这明显是不恰当的。考虑到下面两项 $Q_t$ 与房间的制冷量不直接相关，因此，在图中删除，而上面两项若延用 $Q_t$ 符号，应在前面加负号"-"，表示为"$-Q_t$"。

内热源发热量实例　　　　　　表 5

| | 运行时的发热量 |
|---|---|
| 电脑和显示器 | 150W |
| 激光打印机 | 190W |
| 人、久坐的活动 | 120W |
| 灯光 | 10W/m²（敞开式的区域）|

然而，这个简化的方法常常导致系统过大（或大号系统），因此，没有比供暖与通风工程师更加精确的分析更重要的了。

能够估算出一个房间大概的冷负荷，并将其与多种供冷系统的容量做对比，这对一个建筑的初步设计中供冷系统的最初选择有帮助。表 5 给出了一些办公室中常见冷负荷的近似值。

在前述对比过程中，设计者必须核查到一个程度，一个使用简化冷负荷计算方法算出的需求也能够通过蓄热体和通风系统来满足的程度。因为，这通常会导向一个更经济有效的系统规模。

## 1.3 补进需求

### 1.3.1 回避原则

为保持建筑的能量需求尽可能的低，除了供暖、供冷和通风所需的最终能量需求之外，一个设计者必须同时考虑用户的特殊需求（例如

> 示例：
> 对于一个建筑面积为 10m² 的办公单元，供冷负荷可粗略地按照如下方法估算：
> 根据表5，建筑面积为 10m² 的办公室内部的峰值冷负荷估计值为 560W（150+190+120+100）或 56W/m²。然而，来自某些设备的冷负荷并非恒定，如打印机或照明设备；因此，这些估计值可能被认为有些高。例如，当打印机或照明的运行时间减少 50% 时，室内冷负荷为 41.5W/m²。此外，通过立面进入建筑的太阳辐射也必须加以考虑。

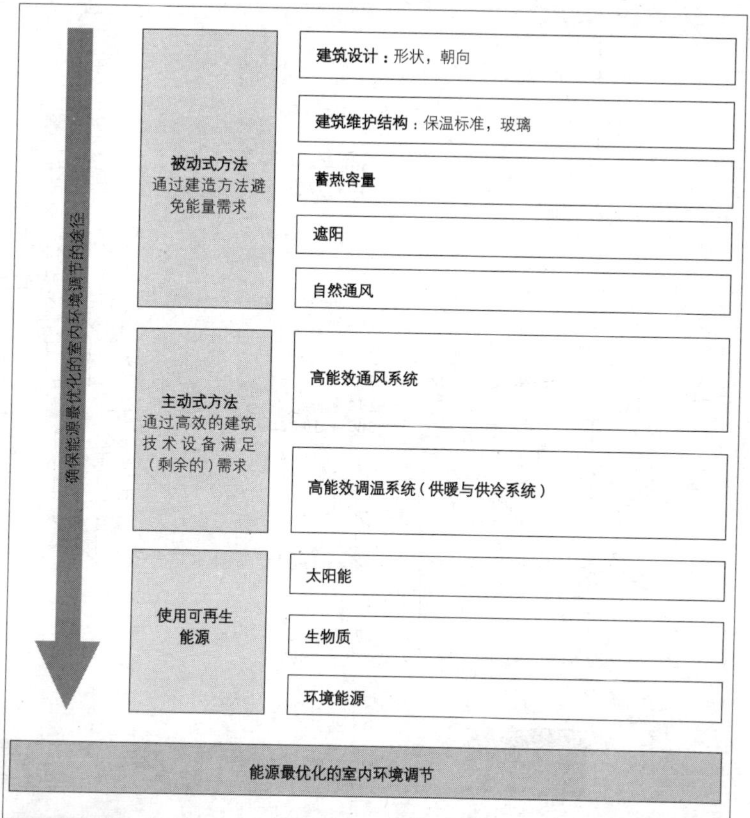
确保能源最优化的室内环境调节的途径
被动式方法
通过建造方法避免能量需求
建筑设计：形状，朝向
建筑维护结构：保温标准，玻璃
蓄热容量
遮阳
自然通风
主动式方法
通过高效的建筑技术设备满足（剩余的）需求
高能效通风系统
高能效调温系统（供暖与供冷系统）
使用可再生能源
太阳能
生物质
环境能源
能源最优化的室内环境调节

图12：
保证能量最优化的室内环境调节方法

电脑、机器等所需的电能）。为避免能量需求（首先要考虑）或提供所需的能量，可能需要混用不同的被动式与主动式方法，参见图 12。

### 1.3.2 被动式方法

使用被动式方法可在不消耗显著能耗的情况下影响建筑的得热与热损耗。当然，这并非总能实现需要达到的室内条件，但是被动式方法的效果符合需求的方向，而且采用主动式方法可进一步削减需求（如能量需求）。目前，存在多种被动式方法可供选择，它们在某种程度上依赖于一栋建筑的基本结构，因此应在规划过程的最初阶段予以考虑。

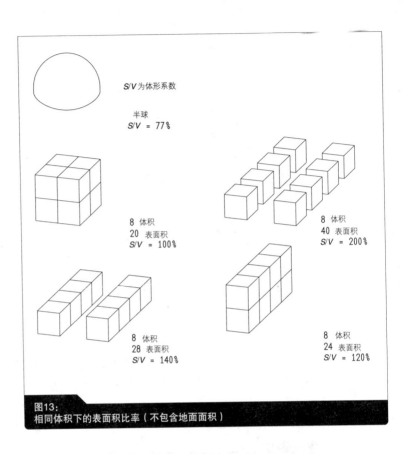

图13:
相同体积下的表面积比率（不包含地面面积）

建筑形状　　一栋建筑通过它的表面（或维护结构）与外界相互作用。这个热传递表面的尺寸也决定了热传递的量。因此，对于建筑能量平衡来说，建筑外形是一个非常重要的设计参数。建筑容积的能量特征被称为体形系数 $S/V$，它是建筑热传递表面的面积与该表面所封闭的建筑体积的比值（参见图13）。

正常情况下，多个参数在一栋建筑的设计阶段已被确定，例如房间高度、房间的日照深度或者功能需求。这就限制了之后通过建筑形状调整来获得能量优化的范围。最终的建筑方案经常包含一份带有多个独立目标组件（独立的建筑）的布局图和具体的布局规划（末端链接引线块等），这会导致比较大的表面积及与之对应的能量效应。谨记，紧凑的形状和最小数量的独立建筑造就更好的能量解决方案。

**定位及气候区域**　建筑中的气候分区与朝向能够使日照得热量最大化，进而具备节能与舒适并举的优势。在一栋以高供暖能量需求为特点的住宅建筑中（例如在欧洲北部），对于最高温度需求的房间（如起居室或客厅），应该使它们最大的玻璃区域面向太阳，以此来使获得的日照收益最大化；而较低温度需求的房间（如卧室），可以背朝太阳。一个内热源发热量很高的房间（如一个有很高使用率的会议室）也应该尽可能不朝向太阳，以避免额外的过热和由太阳辐射的进入引起的各种冷负荷。此外，未分配使用或者对温度和舒适性要求较低的区域（例如流通区域），也可以临近主要使用区域布置，用作室内和室外气候区域的一个缓冲区域。

**保温标准**　保温是建筑内部和外部空间的屏障。因此，保温标准是衡量建筑热传递维护结构质量的标准。在温和的气候区域，因为室外温度在一天或一年的过程中波动幅度较小，所以保温通常就不那么重要了，但也有例外。在炎热与寒冷地区，普遍认为良好的保温标准可以有效地避开供暖与供冷负荷。

在这里，透明部件（窗户、天窗等）特别重要，因为窗户的保温标准通常远低于不透明物体的保温标准。取决于窗的质量和建筑的位置，通过大窗获取日照得热的优势可能会被其在冬天所致热损失的增加抵消掉。

**热存储**　每一种材料都具备一些储存热量并随时间的推移再释放热量的能力。存储与随时间推移释放热量的量值取决于房间维护结构（墙、地板、顶棚）的材料。例如，混凝土与砌体块能够比木材或石膏板存储更多的热量。这以房间或建筑的"热质量"来表示，参见图 14。

尽管热质量对得热与热损失没有直接影响，但其对热负荷与冷负荷非常重要，并因此对建筑在任何特定时间里的能量需求也非常重要：大的热质量能够存储并释放更多来自室内空气的热量，从而避免供热与供冷需求。通常，大的热质量也可以提高舒适度，因为表面温度随着空气温度的升高而上升，且当温度降低时，表面温度需要更长时间才随之下降。

**遮阳**　对室内环境调节，建筑的遮阳是最重要的被动式方法之一。如前所述，进入建筑的能量显著影响建筑的能量需求。因此，遮阳必须尽可能地满足两个相反的目标：在寒冷的天气使进入室内的日照能量最大化，在炎热的天气避免过热。

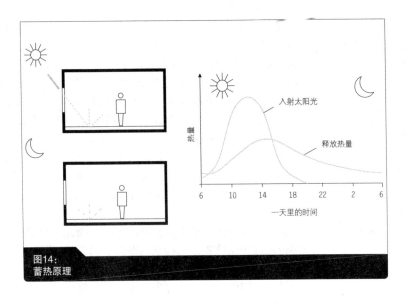

图14：
蓄热原理

遮阳可以分为固定式和可调式两类。固定方法如窗尺寸、朝向、其他建筑的遮蔽（项目中的建筑或附近的建筑、雨棚、树木等等），由于通过所采用的玻璃投射的日照能量总量无法根据季节或每天的时间的变化而改变，这使其与可调式相比处于劣势。参见图15。

提示：

设计窗户以及日照屏蔽元素的其他信息可以参考本套丛书中由罗兰·克里普纳（Roland Krippner）和弗洛里安·穆索（Florian Musso）编著的《建筑立面开洞》一书（中国建筑工业出版社，2013年，征订号：24796）。

建议：

进入室内的太阳辐射能够在一年中的任何时候造成紧急情况，其取决于具体的应用。这种情况经常发生，例如，在内部负荷很高的现代办公大楼。即便在温带气候区（如在德国），与很好的保温标准相结合后，建筑在一年中需要供暖的日子也没几天。在这种情况下，检查夏季是否会发生过热十分重要，例如在使用了大面积的玻璃窗时。

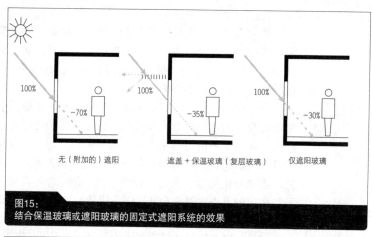

**图15：**
结合保温玻璃或遮阳玻璃的固定式遮阳系统的效果

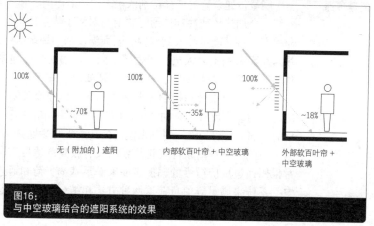

**图16：**
与中空玻璃结合的遮阳系统的效果

可调式方法（软百叶窗帘或卷帘等）允许用户短期影响进入室内的日照辐射量。通过这个方法，就能够尽可能地降低冬季里的供暖能量需求，同时避免夏季里的过热（以及供冷能量需求）问题。因此，与固定式相比，应首选可调式方法。图16展示了可调式遮阳系统的效果。

**自然通风**

自然通风也被认为是一种被动式室内空气调节的方法。自然通风不但可以在不使用能量的情况下满足建筑对新风的需求，而且也能够将暖空气导出建筑并减少冷负荷。关于自然通风概念设计的进一步思考，以及自然通风的局限性，将在"通风系统"一章中展开更详细的描述。

### 1.3.3 主动式方法

当一栋建筑的能量需求不能够通过被动式方法来满足时，必须采用通过引入能量的主动式方法来满足。采用主动式系统也能带来优势，尽管或许在其他领域。因此，每一个概念必须被视为一个整体，包括其所有的组件与能源。

**效率提高**

当考虑为一栋建筑提供剩余能量需求时，所使用的组件都能够尽可能高效的工作是非常重要的。设计者必须确保能量在供应、分配与传递到室内的过程中所涉及的损失越少越好。有众多的不同组件可为此所用，每一个都有其自身的优势和劣势。然而，这些组件之间不能自由结合。因此，必须采用一种综合的设计方法，以便使最初做出的不当决策不会妨碍潜在优势的发展。图17给出了一个如何提高供热系统效率的例子。

**能量传递**

能量传递系统的选择特别依赖于"辐射比例"和"可控性"。

一个加热体或冷却体通过辐射与对流（热量被空气"带走"）复合的方式将其能量释放到室内。表面温度强烈地影响室内操作温度，并因此而影响室内人员的舒适感；因此，辐射热传递系统一般是有利的。参见"设计原理"一章，"舒适要求"一节。

一个传递系统对控制量变化（例如，打开或关闭一个阀门）作出反应的速度反映了它的"可控性"。如果房间的用途与热负荷是变化的，那么"可控性"对其来说是极其重要的参数。

因此，对室内条件变化频繁以及有较高舒适性需求的房间，常常采用通过反应缓慢的辐射表面来满足基本负荷和通过反应快速的空气系统来满足峰值负荷这两种方法相结合的方法。这种结合使它

> 示例：
> 冬季，通过一个供暖系统使一个会议室的室内温度保持在设计范围。当会议开始时，室内坐满了人，所有人都散发出热量，这就意味着已经不需要额外的供暖了。在这种情况下，供暖系统必须能够立即降低它的输出，以避免过热和不必要的能量浪费。

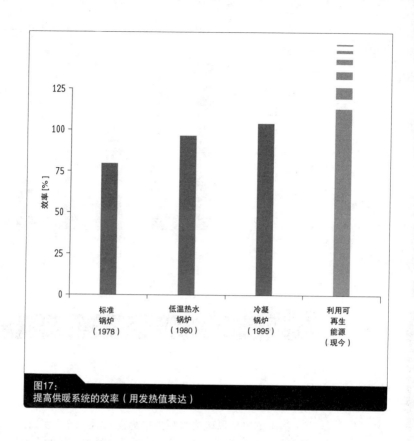

图17：
提高供暖系统的效率（用发热值表达）

们的不足相互抵消，形成一种强大、高效的系统。诸如此类的解决方案的安装成本相对较高。参见"调温系统"一章，"热与冷的传递"一节。

分配　　将能量由其发生器输送并分配到其中转站对整体概念也很重要。首先要考虑的是，每当通过管道或者风管输送水或气体时，都会产生能量的损失。这些损失一部分源自流体与管道内壁的摩擦，另一部分源自温度损失，所以当传输介质抵达中转站时其温度通常会降低[*]。因此，为降低热损失，应尽可能保持管道长度越短越好，且管道保温应尽可能好。

---

[*]译者注：对应供暖工况

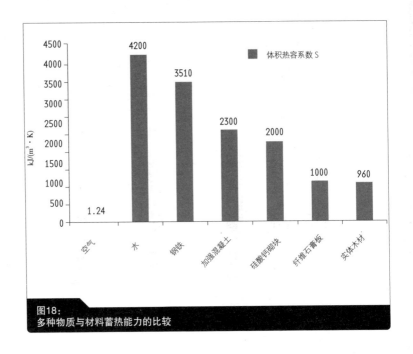

**图18:**
**多种物质与材料蓄热能力的比较**

**选择传输介质**

传输介质的选择至关重要。水和其他液体存储热量的能力远远超过空气，参见图18。

当考虑电动风机或水泵的电力需求时，可证实以水作为能量传输介质比空气更加高效。当考虑输送热量或冷量的附加能量需求时，水操作系统比全空气操作系统更为可取。空气操作系统也是可行的，例如，如果需要满足的供暖或供冷负荷较小且一个通风系统因其他原因必须使用时。参见"调温系统"一章，"热量与冷量"的分配一节。

**能源供应**

在决定建筑的能源供给时，必须考虑能源来源的问题。除实际需求（热、冷、电力等等）之外，这一决定首先受到能源利用的可能性的影响。例如：对于化石能源，问题主要围绕着接入既有服务（燃气、电力、集中供热等）的能力；对于可再生能源，使用他们的机会很重要（太阳辐射、地热、生物质等）；对于一些能源（如石油、木材等），就地储存它们的能力必须给予考虑。参见"调温系统"一章，"能量供给"一节。

和能源的可得性一样，负荷曲线（能量需求变化与时间的关系）在能源的选择上至关重要：由于气候条件或使用功能以及季节或一天中时间的差异，需求可能剧烈变化。这些负荷需求的波动对一个系统的整体效率具有相当大的影响。

关于这点，需要注意能量供给与能量需求的同时性，以及任何必要的能量存储方法。例如，提供热水的太阳能集热系统，仅在有太阳辐射的情况下工作，也就是白天；生活热水需求符合典型使用曲线，即早晨和晚上的峰值用量比一天内的其他时段高很多；因此，白天所产生的热量必须被存储在一个缓冲区，以便在峰值时段内使用。

### 1.3.4 化石能源与可再生能源

虽然提供新风的机械通风系统通常选择电能作为能源，但是经常需要对生成热与冷的能源进行选择，所有这些能源都能够用于多种不同的产能系统。如下分别介绍化石燃料和可再生能源。

化石燃料

化石燃料（如石油、汽油、煤）是在地下或地球表面经过极其漫长的生物和物理过程形成的，因此其在短时间内不可能再生。地球的现有供给是不可再生的，且其存储量也是有限的。这些能源均以碳为基础，经燃烧后以 $CO_2$ 的形式释放到大气中，因此是引起全球变暖的一个重要原因。

在过去，几乎所有的建筑通过化石燃料获取能量，因此必要的技术被广泛开发并可以达到较高的运行效率。尽管有这些积极的发展，供给建筑的能量仍是全球 $CO_2$ 排放的主要来源，这就要求扩大可再生能源的应用。

可再生能源

在人类尺度上可再生能源是用不完的，因此是可持续的：它们可以在不持续破坏环境的情况下开采利用。

> 提示：
> 短语"可持续"来自于林业，它描述了这样一个原则：在一个区域进行采伐，采伐树木的数量与该区域每年可恢复树木的数量一致。采用更一般的表述方法，这个术语指的是仅在可以长期保存其重要特征的方式下使用一个自然系统。

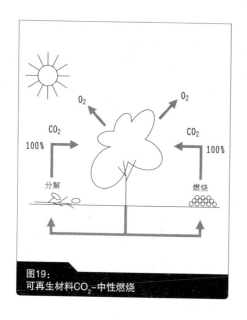

图19:
可再生材料$CO_2$-中性燃烧

从这个意义上讲,太阳能(生成电或热)、水力发电、风能、地热和生物能源(生物质,如树木;沼气,如来自污水处理厂的甲烷),被认为是可再生能源。尽管生物质与沼气的燃烧也通过燃烧过程释放 $CO_2$ 到大气中,这些 $CO_2$ 是在最近的过去伴随植物的生长被其从大气中吸纳的,并且也可能通过植物的自然腐烂过程再次释放。因此生物能源的燃烧被表述为 $CO_2$- 中性。参见图 19。

因此,通常情况下关键问题并非是 $CO_2$ 排放,而是避免 $CO_2$ 排放无法实现和非燃烧所致的大气污染。

在建筑中使用可再生能源的技术体系最近取得了相当大的进展,并且若干年的经验已经证明它们是可靠的。

### 1.3.5 环境影响

很长一段时间,对一栋建筑能量需求的考虑仅仅局限于发生在此建筑中的需求(如供暖能耗)。为评估对环境的影响,这种看待问题的方式就不够充分了,因为建筑中的设备的能量损耗(例如,在锅炉加热水的过程与将被加热的水从锅炉输送到散热器过程中的能量损耗)并没有被考虑。另一方面,能量会在从其发生点到其在建

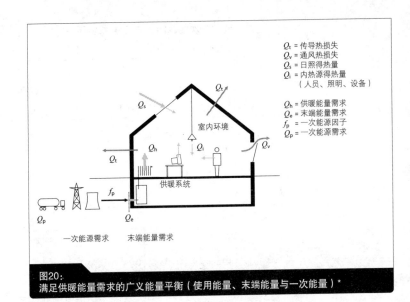

图20:
满足供暖能量需求的广义能量平衡（使用能量、末端能量与一次能量）*

筑中的供应点的途中损失。因此，我们现在要对建筑中的使用能量、建筑边界处的末端能量需求与一次能量需求加以区别；其中，一次能量需求描述的是对自然界创造能源的需求。参见图20。

**一次能源因子**

满足末端能量需求所需的能量支出是使用一次能源因子来定义的，这些能量支出包括能源提取的准备环节（物质条件以及附加能量）、能源加工、能量转换、能量传输与分配等过程的能量消耗，参见表6。

电力通常通过几种方法产生，例如在煤电站、水力发电站或核电站。针对"电力结构"的一次能源因子可以通过发电使用的化石燃料、核燃料与可再生能源的比例计算得出。这个因子因国家的不同而不同，因此可以用来判断电力的使用效果。

**$CO_2$ 排放因子**

**g/kWh$_{end}$**

采用与一次能源因子相似的方法可计算出 $CO_2$ 排放因子，以估算每千瓦时（kWh）末端能量消耗对应的温室气体排放量（单位为克,g）。这个因子的单位可表示为 [g/kWh$_{end}$]。除 $CO_2$ 之外，这个因子也考虑其他污染物的排放量，并以"$CO_2$ 当量"来概括它们产生的温室效应。参见表6。将一栋建筑的能量需求与该因子相乘，就可以计算出供给建筑的能量对全球变暖的影响。

---

\* 译者注：本图删除了 35 页 Fig.20 中的两处 $Q_i$，原因同 101 页图 9 所做改动

## 欧洲一次能源因子与 $CO_2$ 排放因子
### （基于 DIN-V-18599-1:2007-02）

表6

| 能源 | | 一次能源因子（不可再生比例）[$kWh_{prim}/kWh_{end}$] | $CO_2$ 排放因子（$CO_2$-当量）[$g/kWh_{end}$] |
|---|---|---|---|
| 燃料 | 石油 EL | 1.1 | 303 |
| | 天然气 H | 1.1 | 249 |
| | 液化气 | 1.1 | 263 |
| | 煤 | 1.1 | 439 |
| | 褐煤 | 1.2 | 452 |
| | 木材 | 0.2 | 42 |
| 集中供热（70%）热电联产供应 | 化石燃料 | 0.7 | 217 |
| | 可再生燃料 | 0.0 | |
| 集中供热 供热厂供应 | 化石燃料 | 1.3 | 408 |
| | 可再生燃料 | 0.1 | |
| 电 | 混合电力 | 2.7 | 647 |
| 环境能源 | 太阳能、环境热能 | 0.0 | |

图 21 给出了能源选择对一栋建筑一次能量需求与 $CO_2$ 当量排放量影响的一个例子。

从图 21 所示内容可知，如果使用电力来满足一个居住空间每年 $50kWh/m^2$ 的末端能量需求，那么所需的一次能量几乎是末端能量需求[*]的 3 倍。采用燃气或煤时，所需的一次能量仅为末端能量需求的

> 提示：
> 一次能量因子的计量单位是 [$kWh_{prim}/kWh_{end}$]。该因子表示每提供 1kWh 末端能量（例如电能或热能）需要多少 kWh 的一次能量（例如一种具体的能源的用量，如煤）。

[*]译者注：原文此处为环境能量，但环境能量对应的一次能源能量为零，不能用于比较；加之环境能量数值与末端能量需求相等，因此译文中在此处采用"末端能量需求"

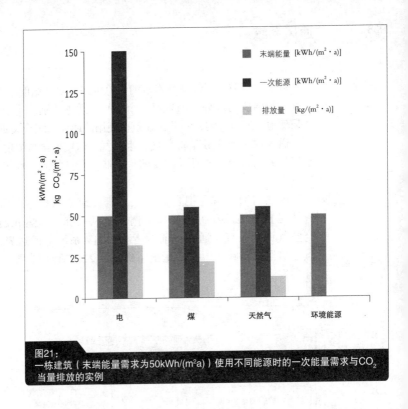

图21:
一栋建筑（末端能量需求为50kWh/(m²a)）使用不同能源时的一次能量需求与$CO_2$当量排放的实例

1.1倍,即仅高10%。另一方面,环境能源几乎就是一次能源中性(一次能源的需求量几乎为零)。上述分析同样适用于$CO_2$排放。

在当地环境允许的情况下,应该尽量避免使用电力及化石燃料。

## 通风系统

从上一章节关于舒适性需求的讨论可知,空气温度、空气流速、湿度、清洁度与成分等因素对一个房间舒适感特别地重要。这些因素强烈地受通风影响,因此,建筑所使用的通风系统至关重要。

通风的主要任务是将臭气、水蒸气、$CO_2$与含高浓度污染物的空气从室内排出,并且最重要的是创造一个良好的、均匀的室内热环境。

供热与通风行业把自然通风和机械通风从根本上区别开来,流体是自然通风和机械通风(例如通风系统、加热和通风系统,与空气调节系统,通常把前两者称为HV系统)的边界,但这一点在口语或实践之中并不是很明确。图22展示了通风系统的一种分类方法。

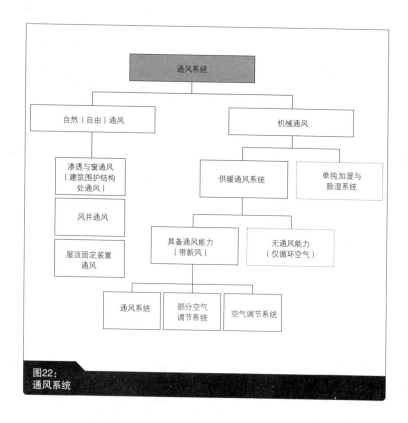

图22:
通风系统

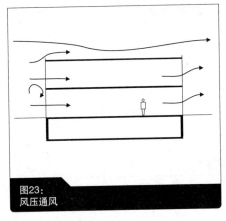

图23:
风压通风

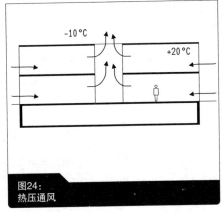

图24:
热压通风

## 2.1 自然通风

自然通风主要依靠建筑表面与其内部的压力差来驱动室内空气流动。这些压力差由风（见图23）或者温度差（热压，见图24）引起。

由图22所示可知，自然通风可以划分为以下三种：
− 渗透通风与窗通风（建筑围护结构通风）
− 风井通风
− 屋顶通风

所需要的空气体积流量主要依赖于天气、室内温度、通风口的布置方式和通风口的气动设计。

### 2.1.1 渗透通风

"渗透通风"描述的是发生在建筑漏风处的空气交换，主要指通过开启的窗及门周围的缝隙的空气交换。这种形式的通风存在诸多问题。

当无风并且温差较小的情况下，这种通风方式无法保障满足卫生要求所需要的最小换气次数。还有，渗透通风往往不受用户影响，持续的不受控的通风会导致热损失的增加，这可能会使建筑严重损坏。为了避免发生前述问题，并允许一定的与需求相匹配的换气率，一栋节能建筑的外墙（包括窗户）必须尽可能地提高气密性，并且不应该有不可控的缝隙或连接点。相应的，新风需求必须需要通过其他通风方法来满足。

**使用窗通风的换气次数**　　　　　表7

| 窗通风类型 | 换气次数 (1/h) |
|---|---|
| 窗、门关闭（仅渗透通风） | 0～0.5 |
| 单面通风，窗倾斜（无板条遮光窗帘） | 0.8～4.0 |
| 单面通风，窗半开 | 5～10 |
| 单面通风，窗全开倾斜[1]（净化通风） | 9～15 |
| 穿堂风（通过相对的窗或门的净化通风） | 可达45 |

[1] 净化通风4分钟实现一次换气

### 2.1.2 窗通风

建筑或房间通常利用窗或者其他常规开口（如风门）进行自然通风。根据需要,这些开口或在短时间内处于开启状态（净化通风）或者长时间开着（持续通风）。对于大多数的建筑，这就是它们常年使用的用以保持室内舒适的通风方式。

在某些气候区的严冬或盛夏,受窗通风引起的热损失或冷负荷过大等问题影响,只能采用间歇的净化通风。

**窗户类型**　　进入和流出滑窗与翻窗的空气量同样大且可调节,这使它们比倾斜式窗更适合自然通风系统,参见图25。

**空气流动特性**　　窗通风的空气流动特性在冬天及夏天是不同的,这种差异取决于室内和室外空气的温度差,参见图26。

**穿堂风**　　打开位于一面墙上的窗仅能提供单面通风。理想情况下,应该通过位于相对的两面墙上的窗之间形成的穿堂风为建筑提供通风。在住宅建筑中,应该有足够的穿堂风,或者至少有横穿一个角落的通风,参见图27。

**换气次数**　　表7给出了多种不同窗通风设置情况下换气次数的粗略推荐值。

窗风或渗透通风可提供的换气次数的预期值波动并且强烈受到空气流速与建筑几何形状的影响。一栋建筑或一个房间能够进行自然通风的尺寸受一些条件的限制。对于一个净空高度为4m的房间,采用单面通风时,房间进深不应该超过房间高度的2.5倍；若采用对流通风,该比值可提高到5,参见图28。

表8给出了自然通风系统设计推荐值的概况。

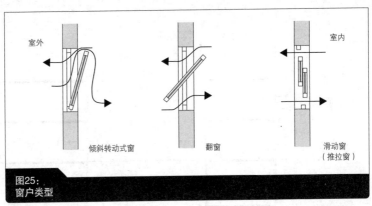

图25：
窗户类型

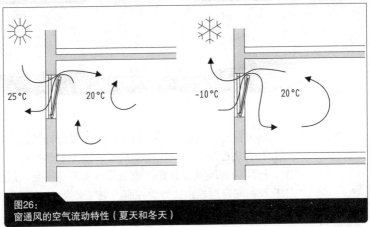

图26：
窗通风的空气流动特性（夏天和冬天）

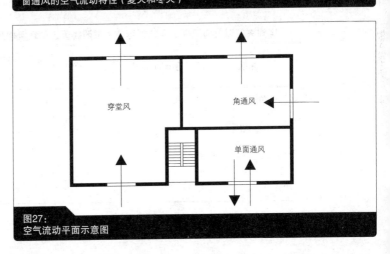

图27：
空气流动平面示意图

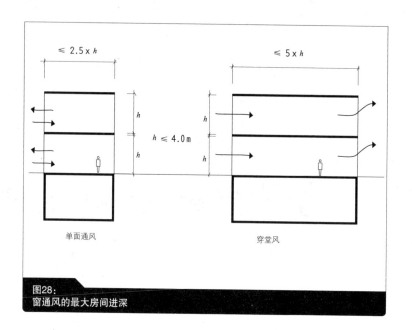

图28:
窗通风的最大房间进深

用户行为　　通过窗通风所能达到的换气次数主要依赖于用户的行为。若用户能够直接控制进入室内的空气流,这无疑是有利条件;但在实际中,房间的通风量往往是严重过量或者严重不足。因此,不能采用用户控制的窗通风来代替受控制的空气交换。

房间深度以及通风横截面的推荐值（依照德国工作场所规则）　表8

| 系统 | 顶棚净高 (h) | 最大房间进深 | 进风与排风截面面积,$cm^2/m^2$ 房屋面积 |
| --- | --- | --- | --- |
| 单面通风 | 4.0m 以下 | $2.5 \times h$ | 200 |
| 穿堂风 | 4.0m 以下 | $5.0 \times h$ | 120 |
| 带屋顶设备与在外墙或在相对两面外墙开口的穿堂风 | 高于 4.0m | $5.0 \times h$ | 80 |

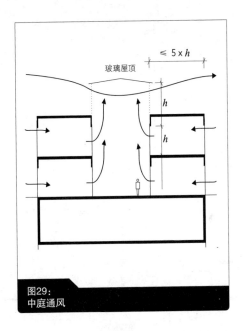

图29：
中庭通风

### 2.1.3 风井通风

风井通风是利用热压原理的一种自然通风形式（参见119页图24），热压的形成涉及井口顶部室内和室外空气的温降与空气绕流井口顶部时产生的吸力效应（参见图30）。

如果这些条件都不存在，例如，在夏天当室外和室内空气温度可能相等或者无风时，这种类型的通风方式在没有风扇补偿下是无

> 建议：
> 　　即使在很深的建筑中，也应该设计天井，为房间提供光线及空气。风井自然通风与烟囱的工作原理一样：被加热的空气上升并通过屋顶的开口逸出，这就产生了一个可吸入邻近房间废气的低压区域。如果天井被玻璃覆盖，则必须提供足够多的垂直气体排出口来防止建筑内部的空气在夏天过热。参见图29。

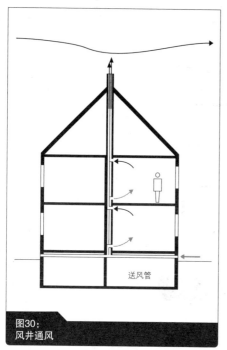

图30:
风井通风

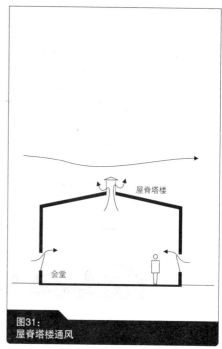

图31:
屋脊塔楼通风

效的。因此,风井通风仅适用于主要需求为排出水蒸气的房间,例如浴室以及厨房。如果这种问题仅在短时间内存在,并且维护结构很厚,那么这种通风方式仍是可以接受的,因为当有足够的热压时,烟囱效应便会恢复。确保每一个需要通风的房间都有自己的井口非常重要,只有这样,足够的空气才能够每时每刻都在流通,并且能够避免来自其他房间的难闻气体的影响。

### 2.1.4 屋顶装置通风

屋顶装置通风是指发生在装设在建筑屋顶中的固定装置处的自然通风,如屋脊塔楼、屋顶天窗以及相似的排气窗口,参见图31。

这种形式的通风与风井通风相似,主要是基于室内外温度不同引起热压作用原理。屋顶装置通风主要用于屋顶净高接近或超过4m的房间(参见表8),即大礼堂(工业棚屋)。屋顶通风无运行成本,但是会导致冬季的热损失大幅增加。当前,通常采用带热回收的机械通风扇。

## 2.2 机械通风

机械通风系统包括简单通风系统（例如，带风扇的风井通风系统或外墙通风机）和供暖与通风（HV）系统，机械通风系统在中心设备间集中处理空气，然后通过送风系统（空气输送管道、风井）将空气输送到室内（参见118页图22）。

### 2.2.1 供暖与通风系统

供暖与通风系统可能包含或不包含通风。没有通风组件的HV系统仅仅循环室内空气，并不引入新风。这些系统主要用于特定的工业制造过程，在此不予展开讨论。

**包含通风组件的HV系统**

HV系统通风的目的是更新与过滤室内空气。该系统常常需要引入一定比例的新风，并排出废气或污浊的空气，以保证室内空气始终保持清新。如果室内空气没有被臭气或污染物污染，那么部分排风可以通过混合段循环使用，即该部分回风与新风在混合段混合后重新被送入室内。HV系统也可以对空气进行加热、冷却、加湿与除湿等热力处理，相应的可以根据系统对送风的处理方式对系统进行分类：

- 没有空气热力处理能力的通风系统，或仅有一项热力处理能力的（例如，仅有加热能力）的通风系统
- 有两个或三个空气热力处理能力的部分空气调节系统（例如加热与制冷，或者加热、冷却与除湿）
- 拥有四种空气热力处理能力的空气调节系统（加热、制冷、加湿与除湿）

**集中HV系统**

集中供暖与通风（HV）系统往往需要机械装置来处理空气，且当其超过一定尺寸后，就必须被安装在其专属的房间（HV设备间）之内。

> **提示：**
> 热回收设备从排风中吸取热量（或冷量）来加热（或冷却）新风，因此，可节约一部分热量或冷量。若不用热回收装置，则需要额外提供这部分热量或冷量来调节新风温度。

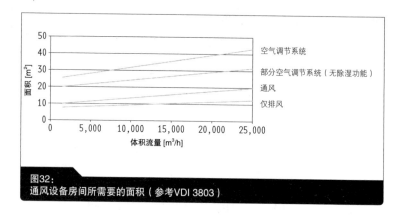

图32:
通风设备房间所需要的面积（参考VDI 3803）

新风负荷必须首先根据房间的预期用途、室外气候与室内环境要求进行计算，参见"设计原理"一章中"确定需求"一节。

如果已知所需的（室外）空气体积流量 $\dot{V}$，再综合考虑所需的空气处理类型，就能够根据图 32 所示的保守数值，确定集中 HV 系统所需要的空间。

### 2.2.2 分配

如果采用中央空调模式，那么空气就需要通过管路系统分配到整个建筑中去，其中的管路系统包括风管、垂直的竖井或水平的地板区。送风系统的设计应该满足通风系统对空间的需求。

通风管道的形状由流动特性与可使用的空间所决定。圆形截面价钱低并且有很好的流动特性，但是与矩形的相比需要更多的空间；

> 通风管件需求的形状及空间

> **提示：**
> 非集中的通风系统是一个替代集中空气调节系统的选择。它能够减少所需的竖井或者风管的面积。使用非集中通风机组，每一个房间所需的新风都能够直接被供给，而且所供给的空气还能够按照个性化需求通过加热或者冷却进行调温（参见"调温系统"一章中"热与冷的分配"一节）。

> **提示：**
> 一个负责送风与排风的通风系统需要独立的送风与排风管道，这在分配空间时必须加以考虑。送风管与排风管不能相互交叉，同时，悬挂的顶棚内必须能够容纳两种管道（一种管道在另一种管道的上面），否则就可能在特定环境下出现顶棚高度显著降低的问题。当最小顶棚高度必须遵守时，建筑成本可能会因此而显著增加。

## 示例：

已知一个房间的送风体积流量为1200m³/h，管道中空气流速为3m/s，计算所需的送风管道的横截面积：

$A = \dot{V}/(v \times 3600)$
　　$= 1200m^3/h/(3m/s \times 3600s/h)$
　　$= 0.1111m^2 = 111.11cm^2$

如果使用一个矩形的送风管道，且使用1∶3这样一个节约空间的边长比，则截面的尺寸可以通过如下方法计算：

$b/h = 1/3$

$b = 3 \times h$
$A = b \times h = 3 \times h^2$
$h = \sqrt{(A/3)}$
　$= \sqrt{(1111.11cm^2/3)}$
　$= 19.24cm \approx 20cm$ 为选定的管道高度。
$b = 3 \times h$
　$= 3 \times 20cm = 60cm$

如果选择一个圆形截面的风管，那么它应具备的直径大小约为40cm，相应的，需要更大的安装空间（如图33）。

因此，可以实现边长比从1∶1（正方形）到最大1∶10（平面长方形）的矩形管道的应用更为频繁。为了降低压力损失以及相伴的气体噪音，应该尽量避免管道方向的突变。此外，采用重量大的管道与横截面大的管道（相应的空气流速较低）也可以降低空气噪音。

送风管道中，空气通常采用v=3~5m/s的流速进行输送（但是在住宅建筑中速度为v=1.5m/s，或者更低）。通过上述假设与所需的空气体积，就能够通过以下公式计算出一个风管所需的横截面积的保守值：

$$A \ \frac{\dot{V}}{v \times 3600} \ [m^2]$$

式中：$A$ 为管道横截面积，$m^2$；$\dot{V}$ 为空气体积流量（空气容积），$m^3/h$；$v$ 为风管中的空气流速，m/s（$= 3600 \times m/h$）。

**保温隔热**

通风管道必须采用保温隔热来降低噪音、减少供暖时的热损失与避免风管结露。通常使用厚度约为5cm的隔热层对风管进行保温。在考虑安装风管所需的空间时，这个厚度必须被加到风管的尺寸上。此外，还需在风管周围附加5~10cm的工作空间，参见图33。

**消防安全以及隔音**

另外，通风管道必须结合特定的措施来确保消防安全与隔音要求。这是因为通风管道经常穿越防火分区与不同的使用区域。与通风管道相比，适用的防火阀与消声器的外部尺寸通常更大一些，因此在设计中必须考虑这一点，以为这些部件的安装与维护预留出足够的空间。

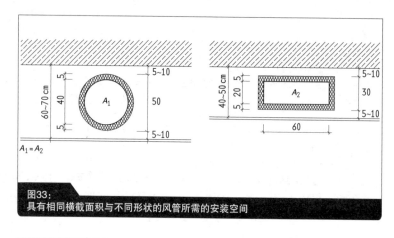

**图33：**
具有相同横截面积与不同形状的风管所需的安装空间

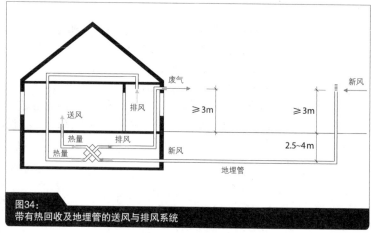

**图34：**
带有热回收及地埋管的送风与排风系统

**废气管道及新风管道**

室外空气引入口与废气排放口都必须加以保护，以防止雨水、鸟与昆虫等进入，并且风口与地面之间的距离至少为3m，参见图34。通过一个埋设深度为2.5~4m的地埋管将新风引入建筑是一种有效的节能手段。空气通过温度相对恒定的土壤，在夏天可被预冷，在冬天可被预热，参见"调温系统"一章中"能源供应"一节。

**热量回收**

通过换热器可将排风中的热量回收并用于加热引入的新风。排风的气流与送入的冷空气在换热器中不掺混的交叉流过，因此回收过程中没有污染物的传递。热回收的效率可达90%以上，但具体数值依赖于热交换器的类型。参见图34。

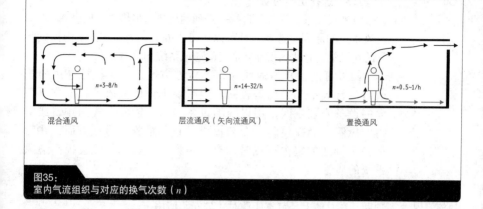

图35:
室内气流组织与对应的换气次数（$n$）

### 2.2.3 室内气流组织

经温湿度调节后的空气的配送方式对室内的舒适性至关重要。室内气流组织的基本途径有三种（参见图35）：
- 混合通风
- 层流通风（矢向流通风）
- 置换通风

**混合通风**　　混合通风是将空气引入室内最常用的方法。送风以一个相对高的流速经顶棚或墙壁上的风口吹入室内，并与室内的静止气体混合。

**矢向流通风**　　矢向流通风是有特殊用途的房间所采用的一种通风方式。供给的空气通过整面墙（或顶棚）引入，并且通过与之正对的面排出。其典型应用为手术室与洁净室，为了制造一个极其清洁的环境，所使用的通风方法必须确保送风与室内空气不发生掺混。

**置换通风**　　置换通风是公认的一种可实现节能与舒适性并举的特殊通风方式。温度比室内空气低约 2~3 K 的新风，以很低的流速（<0.2m/s）贴近地面引入室内。新风在室内贴近地面的高度扩散，进而形成一个新风池。室内的热源（如人体或者电脑）会导致新风以对流方式上升，进而能够为每个人提供足够的新风。因此，换气次数可降低至满足卫生保健要求（$n$=0.5~1.0/h）的最小值,相应的能量需求被大大削减。置换通风作用不受房间进深与建筑物体积的影响，因此，其适用于进深很大的建筑与空气需求量大且冷负荷较低（最大约 35W/m²）的会堂，例如电影院、体育馆或办公楼。

P50

### 2.3 选择正确的系统

每种通风系统都有其自身一系列的优势与劣势，参见表9。确定通风系统的原则是选择一种可以用最小的能耗为建筑或室内提供所需的新风量，并达到很高舒适度的通风系统。

如果可能，自然通风应作为首选方案，因为机械通风的建造与运行成本几乎总是较高。同时要牢记机械通风系统需要很大的设备安装空间，用以容纳设备、用于输配空气到建筑或室内的风管、确保消防安全和提供隔音的措施。另一方面，在营造一个舒适的室内环境时，充分考虑热回收，往往会提高通风系统的质量。

**使用机械通风的原因**

机械通风仅在有功能性需求（构造的或用户相关的）或者能够预期其在总体上是节省能源时才被安装使用：

- 没有窗的，或者内区房间需要新风供给；
- 高度超过40m的建筑建议安装机械通风系统。对于这种高度的楼房，当打开窗户时，风压及对流能够引起强烈的吹风感；

通风系统的特征　　　　　　　　　　表9

| | | 优点 | 缺点 |
|---|---|---|---|
| 自然通风 | | 不需要能量来驱动或处理空气 | 效果受气候条件影响（风速与温度） |
| | | 安装系统不占用室内空间（无风管、设备间等） | 功能受建筑结构与房间进深影响 |
| | | 初投资与维护费用低 | 冬季热损失（或热损耗）大 |
| | | 与户外环境的最佳关系 | 热回收的可能性低或实施非常困难 |
| | | 高用户接受性（或高用户认可度） | |
| 机械通风 | | 容易调节 | 初投资、运行与维护费用高 |
| | | 能进行热回收 | 增加设备与风管安装所需空间需求 |
| | | 适合全部的空气热力学处理工艺（供热、供冷、加湿与除湿） | 低用户接受性，特别是由缺乏用户干预（或调节）功能所引起 |
| | | 空气被污染情况下能够使用过滤器 | |

- 若不采用特殊构造措施，如采用双层幕墙来抵消这种效果，那么较高的楼层几乎无法采用自然通风；
- 建筑处在一个臭气或噪声负荷或废气排放强度大的位置时，机械通风值得安装；
- 房间进深很深，自然通风不能提供所需的空气交换，需要机械通风。参见122页图28；
- 在剧院、电影院以及其他集会的地方，缺少窗户或者窗户面积相对较小，但这些场所上座率高时人员密度很大，这导致自然通风无法单独满足通风需求；
- 在微生物水平、温度、湿度等方面指定室内空气质量的房间，例如手术室、博物馆或者特殊的生产设施（如微处理器的生产设施等），需要使用机械通风；
- 热负荷很高的房间（例如计算机中心与程控机房），需要冷却系统与机械通风；
- 综上所述，机械通风能够成为节约能源的有效措施。通风系统，特别是那些具有热回收的系统，降低了通风热量损失，是被动式住宅能源概念中不可或缺的组成部分。

## 调温系统

主动调温系统的目的是尽可能高效地使用一种能源，为创造一个舒适的室内环境提供任意所需的附加热量或冷量，并将其引入室内。参见"设计原理"一章中"舒适性需求"一节。

一般而言，一个调温系统可以由一个能量源、一个产生热量或冷量的技术系统、一些储存能量的方法（如果需要）、一个与分配和输送——与用户需求相结合的——关联的控制方法和需要调温的目标（建筑或者房间），参见图36。

一个主动调温系统的所有部件必须通过精确的调整以实现相互匹配，这样系统才能始终以一个高效的方式满足建筑的需求。参见"设计原理"一章中"确定需求"一节。

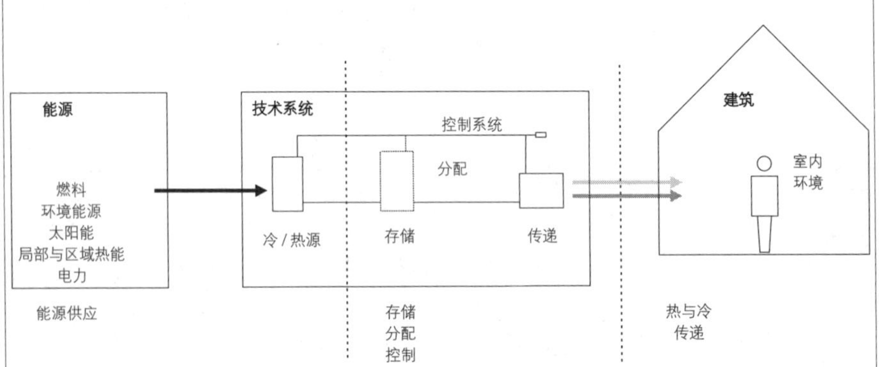

**图36：主动调温系统**

能源
- 燃料
- 环境能源
- 太阳能
- 局部与区域热能
- 电力

能源供应

技术系统
- 冷/热源
- 存储
- 传递
- 控制系统
- 分配

存储
分配
控制

建筑
- 室内环境

热与冷传递

> **提示：**
> 将水由热发生器输送到传输系统的管道被称为"供水管"，将水带回的管道被称为"回水管"。这些管道中水的温度是反映联合制热或制冷机组与传输系统能力的重要参数。

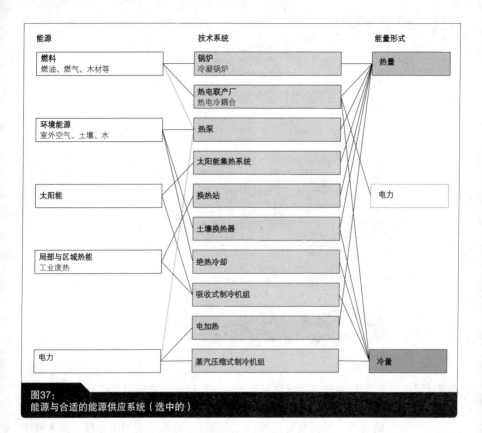

图37:
能源与合适的能源供应系统(选中的)

P53

## 3.1 能源供应

建筑中热量与冷量的供给依赖于可用的能源。组合各类能源和选出的系统进行能源供应的一些可行的途径如图37所示。

### 3.1.1 燃料

石油、天然气或者煤等化石能源主要用于产生室内空气调节所需的热量。

燃烧化石能源产生的环境效应(主要是热量与$CO_2$的排放)致使燃用化石能源受到批判。因此,应优先选用$CO_2$-中性的燃料。参见"设计原理"一章中的"满足需求"一节。

锅炉　　整个世界范围内,供给建筑的热量主要来源于燃烧燃料的中心供暖锅炉。这个过程产生的热量通过热交换器释放到一种热介质(一般是水)中,然后分配到建筑中(集中供暖)。

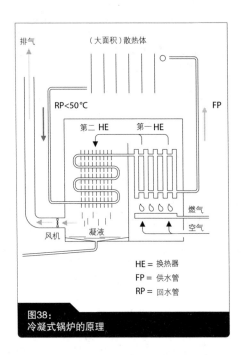

图38:
冷凝式锅炉的原理

冷凝式锅炉　　效率最高的锅炉被称为冷凝式锅炉,它通过从废气中提取额外的热量(指部分潜热)来实现更高的效率。石油和天然气是最常用的燃料,然而在近几年,已经出现一些能够使用木颗粒(由木屑压制成型的小棒状可燃材料)作为燃料的系统。冷凝式锅炉的原理如图 38 所示。冷凝式锅炉最好与供暖散热器结合使用,因为该锅炉需要较低的回水温度。

建议:
　　冷凝式锅炉的系统温度低这一特点使其能够很好地与太阳集热系统相结合。排气温度低要求冷凝式锅炉配套的烟囱内壁衬一层抗凝结膜,并且需要加装排气扇,因为对流效应很弱。

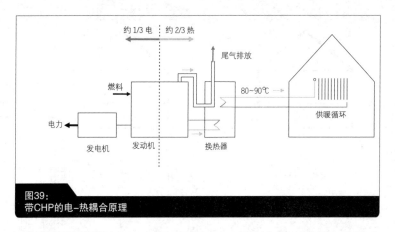

**图39:**
**带CHP的电-热耦合原理**

**电-热耦合/热电联产厂**

除了燃烧化石能源或者生物能源产生热量之外，建筑也能够采用其他过程产生的一部分热量。其中，一个很重要的例子是使用发电过程中产生的热量进行供暖。

与产生热量一样，发电也常常依赖于燃烧过程产生的高温排气。燃烧过程产生的热量的迁移最终转化为两种形式的能量，例如，在一个水操作系统中（以水为工质的热电系统），这两种形式的能量为：电能和热能。因此，这些过程通常被称作"电-热耦合"（PHC）。根据燃料的耗量，这些过程非常高效，因为燃料中所含的大部分能量都能被提取与应用。

PHC的原理在一个大型的供热电站中同样能够被很好地运用，如在中型电厂（例如为一个住宅区供应）或者小型电厂（例如为一栋建筑供应）中。中型或小型的电热厂被称为热电联产厂（CHP），其提供的热量主要以温度为80~90℃的热水供出（参见图39）。最常用的燃料是天然气与供暖用轻油，但沼气与生物燃料（例如菜籽油）也可以使用。

结合吸收式制冷设备，PHC产生的热量能够像太阳能产生的热量一样被用来制冷（参见后续3.1.3：太阳能）。因此，这些过程被称为"电-热-冷耦合"（PHCC）系统[*]。

### 3.1.2 环境能源

存在多种应用建筑周围物体的能源潜力来为建筑制热与制冷的可能。除了室外空气与室内排出空气的温度水平之外，土壤不同深

---

[*] 译者注：国内称之为热电冷联产系统，简记为CCHP

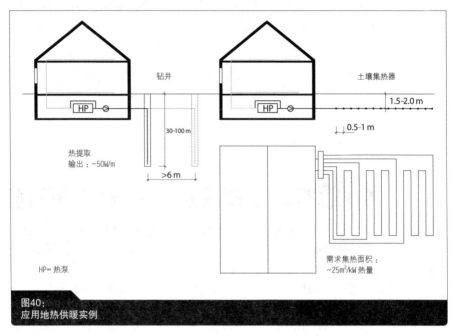

图40:
应用地热供暖实例

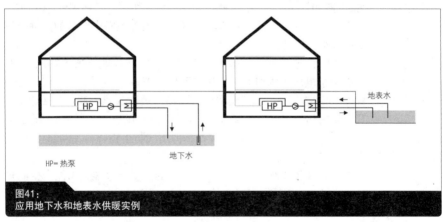

图41:
应用地下水和地表水供暖实例

  度处的温度、地下水的温度以及建筑附近水域水体的温度均相对恒定。环境能源的应用实例参见图 40 与图 41。

  从这些源中获得的能量，在冬季可为建筑供暖，在夏季可为建筑供冷。当环境能源被用于供暖时，通常需要引入附加的能量来提高其温度水平。

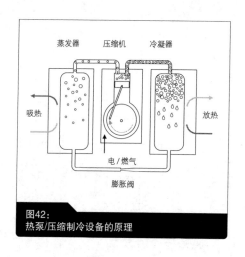

图42：
热泵/压缩制冷设备的原理

**热泵**

提升环境能源的温度水平可通过热泵来实现,热泵的工作原理与冰箱相同,如图 42 所示。环境能源必须通过电或燃气来放大(或提升品位),以将温度提高到期望的水平,保证系统高效运行。

与电能相比,使用燃气具有初级能源的优势(参见"设计原理"一章中的"满足需求"一节)。

**土壤热交换器**

土壤热能("地热能源")的应用主要通过地热收集器或者与热泵相结合的钻井。地热能也可以用来对室外空气进行简单的预处理,即在供给的新风进入建筑之前,先令其穿过一个空气–土壤热交换器。如此一来,新风在冬天可被预热,在夏天可被预冷,进而有利于减少制热与制冷的能量需求,参见图43。

---

**建议：**

热源的温度越高或者热源与供热系统之间的温差越小,热泵的工作效率越高。热泵供热最适合于系统温度低且加热面积较大的系统,如地板供暖。

**提示：**

气体流过土壤热交换器管道会产生附加流动阻力,因此,需要采用机械通风装置。

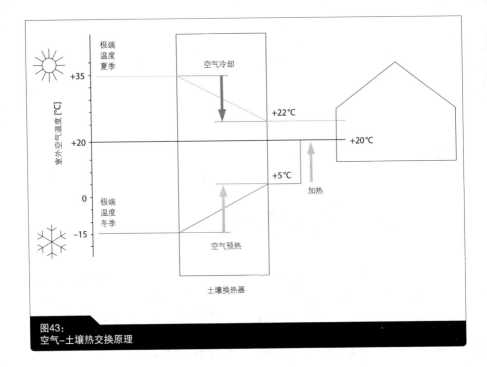

**图43：**
**空气–土壤热交换原理**

绝热冷却　　　除上述利用地下水和地表水作为自然能源用于制热与制冷之外，也可以直接用水来为建筑制冷。与人体皮肤出汗降温的原理相似，水的蒸发也能从空气中提取能量，进而冷却空气。这种效果也被称为绝热冷却，它可以直接在建筑或室内创造，或者间接通过蒸发冷却创造。用直接绝热冷却，所供给的空气越过开阔的水表面或装置，特别是在机械通风系统中，水以极细的雾状喷射入空气，这使其不会滴落到地面，而会以水蒸气的形式进入空气，进而冷却供给室内的空气。这种绝热冷却方式的缺点是送风的湿度会不断增加(参见"设计原理"一章，"舒适性需求"一节中的"热舒适")。这使直接绝热冷却供冷系统最适合在干、热气候区域使用。

　　将蒸发冷却原理与一个热交换器相结合，可在机械通风系统实现降温但不增加湿度的处理工况，这通常被称为间接绝热冷却。因此，可通过蒸发冷却来冷却室内排风，被冷却的排风通过转轮式热交换器排出建筑，而热交换器高效地将排风的冷量传递给反向旋转的新风气流，参见图44。

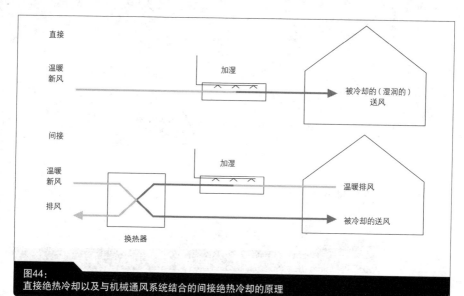

图44：
直接绝热冷却以及与机械通风系统结合的间接绝热冷却的原理

图45：
应用太阳能制热的案例

### 3.1.3 太阳能

太阳能集热系统将太阳辐射转化为热能，这部分热能通常用来为建筑提供热量。

**太阳能集热系统**

通过合适的集热系统，可实现太阳能这一主要用于供暖与制备生活热水的热源的主动利用。这些系统将太阳辐射转化为热量，并

139

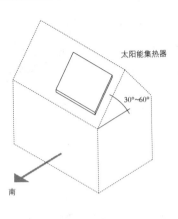

**图46:**
**用于提供背景供暖与生活热水的太阳能集热器的适宜方向**

通过一种热输运介质（添加防冻剂的水）将热量带到其使用场所。太阳能集热系统有两种主要类型：一种使用平板型集热器，另外一种使用真空管型集热器。后一种在制造技术上更为复杂，因此成本更高，但效率也更高。

如果太阳能系统仅仅提供热水，那么太阳热能能够直接进入热水存储罐。提供热水所需能量的不足部分，常常通过联合一个带有或多或少传统特色的锅炉来满足。太阳热能系统也可以用来为具有良好保温的房屋提供背景供暖。参见图45。

太阳采集器常常被放置在屋顶。最优的排列取决于当地太阳的方位与集热系统的使用季节（参见图46）。集热器的安装方向与倾斜角这两个参数出现小偏差是允许的，因为其引起的得热量衰减非常有限。

> **重要：**
> 太阳能集热系统不应该与太阳能电池相混淆！太阳能电池主要安装在屋顶，但是仅仅产生电能。电能不但能够被储存在电池中（岛上的解决方案）或者直接在建筑内使用，而且可以在条件允许的情况下并入公共电网并得到一些经济补偿。

**图47:**
应用太阳能供冷的案例

**吸附式制冷设备**

　　吸附式制冷机组的工作原理与热泵相似，均是通过压缩与膨胀一种介质来实现制冷。其以热源代替电能或燃气，提供驱动制冷循环所需的附加能量。吸附这一化学过程产生的是冷量而不是热量。

　　当有温度在80~160℃的免费热量可利用时，如生产过程排放的废热（生产车间、热电耦合厂等）或可从环境中获取（太阳能集热器、温泉等），采用热能驱动的吸附式制冷机组具有优势。从运行的经济性来看，组合热电联产厂（CHP）和吸附式制冷机组形成的电－热－冷耦合（PHCC）系统是完美的，因为这一套系统不但能够满足制热与制冷的能量需求，而且能够全年运行。

### 3.1.4 局部和区域热量

　　区域热量在集中供热厂或者热电厂产生，也可以在分散的热电联产厂产生，后者被称为局部热量。通过适当的管网系统与换热站，局部与区域热量就能够被应用在偏远的建筑中。利用热电联产（CHP）的原理产生热量是一种维护成本低且环境友好型的方法。参见"设计原理"一章，"满足需求"一节中的"环境影响"。

**换热站**　　使用局部热量或区域热量则不需要在建筑中使用任何热量生产设备，相应地，也就不需要相关的配套设施，例如供暖设备房间、废气站或者燃料储存设施。热量通常是以热水或蒸汽的形式通过保温的管道供给到建筑，然后通过一个换热站（热交换器）转移到建筑的供暖或热水系统中。

**工业余热**　　在某些情况下，能源密集型工业过程（炼钢、化学工业等）的余热也可以通过局部或区域供热管网输送到建筑中。不然，这些能量将会被释放到环境中。这种选择同时也可以提高工业过程的能量效率。

### 3.1.5　电力

原理上，电力也能用于制热与制冷。然而，这种选择应该尽量避免，特别是当电力来自化石燃料时，这是因为电力的产生已经与热量相关，并且在每一次能量转化时（例如，从煤到热量、热量到电能、电能到热量）都会产生能量损耗。这很容易从每一个区域发电对应的初级能源因子得知。参见"设计原理"一章，"满足需求"一节中的"环境影响"。

**电加热**　　电加热供暖系统仅仅被用于特殊情况。例如既有建筑中的浴室。在这里，全年所需的热负荷很小，且其与集中供热系统的连接可能在技术上不可行或者在经济上不合理。上述情况也适用于采用电锅炉或浸没式电加热器的电加热热水供暖。

电能常常作为驱动热泵制热的辅助能量。电能的应用应该尽可能的低，以避免系统整体成本高与初级能源平衡差。

**压缩式制冷机**　　最常用的制冷机是压缩式制冷机。它们的工作原理与冰箱的工作原理一样，都是通过电来制冷，然后将冷量通过分配系统释放到建筑中。使用适当的能量，这种实用方法能够实现任何预期的温度水平。

如果建筑中有热量和冷量的同步需求（例如，同时为室内供冷与提供生活热水），制冷机所产生的废热就能够被利用。这个废热常常被释放到室外空气中，因此结合压缩机的能量需求（通常为电），这一过程仅需要消耗一小部分电能。

对电能能源的依赖不断增加，给初级能源平衡带来不利影响。因此，应尽可能避免采用压缩式制冷机。

## 3.2 热量与冷量存储

通过热量与冷量的存储,能够解除用户处热量利用依赖于热源处热量生产的偶联。这是十分必要的,特别是为了使用太阳能进行供暖,因为从太阳获得的能量随着天气变化,并且在时间上与能量需求不一致(参见图48)。太阳能可以利用建筑自身的蓄热能力进行储存(参见"设计原理"一章中的"满足需求"一节,被动式方法),但是用户无法调节它的使用。用于提供热水以及背景供暖的太阳能系统中通常有一个热水存储罐,该罐能储存足够建筑使用几天的热水,但需要占据一定的建筑内部空间。

季节性蓄热可用于太阳热能的长期蓄存。季节性蓄热通常要用到水槽,该水槽可以安置于建筑内部或者建筑外部。这样就可以在夏季蓄存热量,以供冬季使用。

从原理上讲,同样可以实现冷量的存储,例如,使用"冰罐"使整个系统能够采用同蓄热系统相似的方法进行优化。特别是太阳能,太阳能得热量的峰值与冷负荷的最大值在时间上重合,参见图48。

## 3.3 热量与冷量分配

### 3.3.1 集中系统

连接热源或冷源并将热量或冷量传输到室内末端的系统称为分配系统。例如,热量或冷量大都以集中方式生产,然后以水或空气作为传输介质输送到室内的热交换系统(或简称为末端)。考虑能量

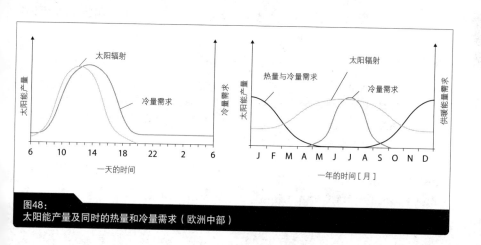

图48:
太阳能产量及同时的热量和冷量需求(欧洲中部)

密度与输送能耗，采用水系统通常更可行，因此，与空气系统相比，在大多数情况下应优先选用水系统（参见"设计原理"一章中的"满足需求"一节）。建筑中最常见的供暖形式是集中热水供暖系统。然而，也有一些设计与经济上的因素有利于空气系统的选用，如下探讨空气系统。

风管管线应该越短越好，并且应该被很好地保温，以减小能量损失。保温同时会防止风管在输运冷媒的过程中结露。

### 3.3.2 非集中式系统

独立供暖机组直接在室内产生与释放热量。有多种类型的独立供暖机组，包括独立的火炉、煤气取暖炉和电暖炉。此外，还有一些特殊形式的供暖系统，本书暂不考虑。

非集中式机械通风系统直接在建筑立面前吸进室外空气，不需要通风设备间以及庞大的管网系统。另一方面，维护大量的小设备成本很高，尽管已通过单元模块化程度的提高与维护的简化来降低设备维护成本。它们通常也能够加热或冷却通风所需的空气（参见"调温系统"一章中的"热量与冷量传递"一节）。

### 3.3.3 调节

除了保温的管道工程、阀门及泵之外，系统中还必须有合适的调节热量及冷量分配的方法。这些控制装置连续不断地使热量和冷量的输出与由天气（室外温度、风、太阳辐射）、内热源与房间用途等变化引起的不断变化的负荷需求相匹配，这些控制装置必须能够

示例：

一个保温隔热等级非常高的建筑，例如被动式房屋，不需要传统的供暖系统（带有供暖锅炉、分配管路与传递热量的散热器）。尽管在物性上以空气作为热传递介质处于劣势，室内所需求的热量也能够通过通风系统来满足，同时提供满足卫生要求所需最小换气次数对应的新风量。

提示：

散热器组是一个例外，其主要用于大型礼堂的供暖，并且需要在较高的供水温度工况下运行。

根据室内与室外的温度实现自动运行，并能够被按照时间进行编程设定。在住宅建筑中，最常用的控制装置包括：散热器恒温控制阀、带有合适传感器的室外与室内自动调温器以及在夜间设定时段和根据温度变化来减小输出的自动控制装置。对于大型建筑或者系统，最好安装一套建筑管理系统（BMS），该系统通过传感器以及中心计算机控制不同房间的加温或降温。

### 3.4 热量与冷量传递

一个合适的热量与冷量的传递系统使用水或者空气作为载体，将热量或冷量从发生器输送并释放到需要调温的房间里。

热量与冷量传递系统可以按照传递方法进行分类，即辐射换热量或者对流换热量的比例、供水温度 $T_f$、额定容量与可控性。参见"设计原理"一章中的"满足需求"一节。

#### 3.4.1 热传递系统

热量可以通过散热器、表面供暖\*，或者通风系统（空气供暖系统）传递到室内。加热元件的结构和布局决定了室内的温度分布，并且对室内舒适度有较大影响。传输到室内的热量应在时间上保持稳定、在水平及垂直方向保持均匀，以实现理想的温度分布。参见表 10 与图 49。

**供暖散热器**

在热水供暖系统中，常规的热传递系统包括散热器、对流散热器与表面供暖。它们应用广泛且易于调节控制。散热器以及类似的供暖部件应该安装在外墙并且靠近玻璃窗的位置，以防止沉降气流的形成（参见图 50）。对流散热器一个典型的缺点是其所需的供水温度较高，这使其很难与太阳热水系统、热泵或者冷凝式锅炉等联合使用。

**表面供暖**

表面供暖系统（地面、墙壁与顶棚供暖）中，隐藏的供暖管道被埋在砂浆层、石膏或者是特殊嵌板中，热量主要以辐射的方式传递到室内，进而创造一个舒适的温度分布（参见图49）。巨大的加热面积允许该系统的供水温度大幅低于其他类型的加热体。对于热负荷特别低以及与低温供暖系统（例如冷凝式锅炉、

---

\*译者注：表面指地面、墙面与顶棚等建筑内表面

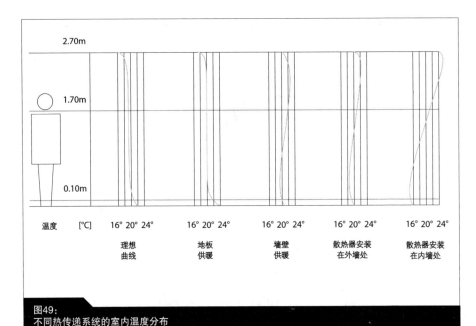

图49：
不同热传递系统的室内温度分布

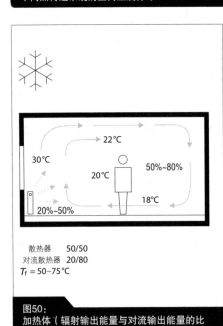

散热器　　　　50/50
对流散热器　　20/80
$T_f = 50\sim75\,°C$

图50：
加热体（辐射输出能量与对流输出能量的比例以及供水温度$T_f$）

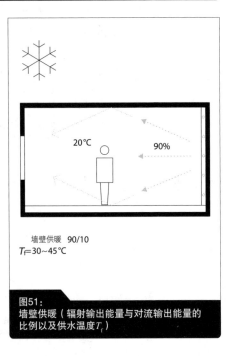

墙壁供暖　90/10
$T_f = 30\sim45\,°C$

图51：
墙壁供暖（辐射输出能量与对流输出能量的比例以及供水温度$T_f$）

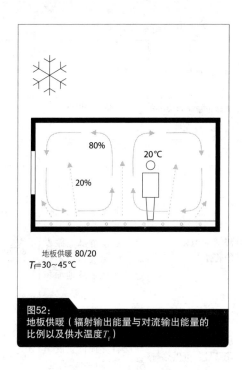

图52:
地板供暖(辐射输出能量与对流输出能量的比例以及供水温度$T_f$)

热泵以及耦合热太阳能系统)联合使用的建筑,表面供暖是一种理想的传递热量的系统。参见图51和图52。

空气加热

原则上,一栋建筑的热负荷也可通过既有的通风系统来满足。供暖所需的热空气可以集中的或局部的制备,热量主要通过电驱动的加热器提供。然而,仅当一个通风系统的需求已因为其他原因业已存在,并且建筑总的热负荷很低(例如建筑具有非常好的保温或者内热源发热量很高)时,这一方案才可行。表10给出了主要的热传递系统的简单概述。

### 3.4.2 冷传递系统

冷量的传递系统与热量传递系统相似,也可以根据传递方法分为对流系统或辐射系统。这一类系统的优点及缺点取决于其对室内操作温度的影响,参见"设计原理"一章中关于"舒适性需求"的描述,其中关于辐射不对称的风险的讨论应该重加以考虑。此外,可控性与所需的供给温度是选择合适的制冷系统的决定性因素。参见151页表11。

热传递系统的特点  表10

| 热传递系统 | 优点 | 缺点 | 输出能量比,辐射/对流 | $T_f[℃]$ |
|---|---|---|---|---|
| 散热器 | 价格低廉,响应时间短,可控性好 | 安装空间,外观 | 50/50 | 50~75 (90) |
| 对流散热器 | 节省空间,响应时间短,可控性好 | 清洗,制造灰尘 | 20/80 | 50~75 (90) |
| 地板辐射供暖 | 舒适的温度分布,隐蔽性 | 可能出现冷空气沉降,相对高的近地面温度可能会使静脉曲张发展,控制缓慢,不适合某些地板或地面覆盖物 | 80/20 | 30~45 |
| 墙壁或顶棚辐射供暖 | 舒适的温度分布,隐蔽性,可供热与供冷 | 控制缓慢,用于供热的墙面前不能放置家具与配件,顶棚供暖所需的足够的顶棚高度,人与加热面必须保持一定距离 | 90/10 | 30~45 |
| 空气供暖 | 通风与供暖结合速效的控制 | 可能出现吹风感,高于49℃会产生灰尘 | 0/100 | 30~49 (90) |

与热量传递相似,利用大型建筑构件的制冷系统对控制的响应速度很慢。空气系统与不包含大型建筑构建的系统对控制的响应迅速,并且对条件的变化反应更加迅速。面积很大的热传递装置与面积小的设备相比更易于拥有适度的(在制冷中:更高的)供给温度,因此,也更加适合与可再生能源联合使用。

对于制冷系统,必须牢记一个特殊点,即温度降低到露点之下的严重性:取决于室内空气的温度与湿度、供冷部件的表面温度,

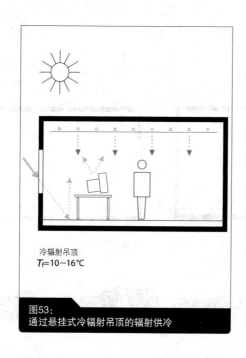

图53：
通过悬挂式冷辐射吊顶的辐射供冷

空气中包含的水蒸气会因为冷表面的温度降至露点温度之下而在其上凝结。生成的冷凝水至少应被收集，更有效的方法是将其排入下水道。如果做不到，冷表面的冷凝水也可以通过临时的提高冷却体的表面温度来处理，当然，这样会降低其冷却能力。

**冷却顶板**　　冷却顶板广泛应用于写字楼与行政办公楼。这些传递系统从一个很大面积的顶棚区域上悬吊下来，并可能覆盖这一区域。参见图53。

内部充满流动冷介质的冷却盘管被悬挂在顶棚上，大约有一半的冷量会以辐射的方式传递，该系统对控制的响应较好。该系统供给温度在10~16℃，且实际上不具备冷凝水排水能力。因此，为了防止结露发生，应该避免较高的湿度水平或在短期内提高冷却水供给温度。特别是与自然通风结合使用时，采用这类系统意味着在温暖、湿润的日子里系统无法充分释放其冷却能力。

**再循环空气供冷**　　利用室内的空气为房间供冷——从室内引出的空气被冷却后再引回到室内（再循环空气系统）是一种广泛使用的解决方案。这类系统通常是分散式的（可能每个房间的系统均是单独的），并且它们对控制响应迅速。它们也能够被房间的使用者控制。在这一系统中，

图54：
再循环空气机组对流供冷

图55：
重力循环供冷的原理

再循环空气供冷
$T_f$=6~10℃

重力循环供冷
$T_f$=6~10℃

**重力循环供冷**

空气能够直接被冷调风器（位于悬吊在顶棚上的设备的排气口处）冷却。在这种冷却方式中，热量以纯对流换热方式传递，冷调风器所需的冷却介质供给温度通常为 6~10℃。参见图 54。

在没有风机的情况下，也能够通过重力供冷原理（或者下向通风制冷）实现完全的循环空气供冷。在重力供冷中，空气的流动由冷空气比热空气重引起。室内空气通过顶棚附近（这里是室内空气温度最高的地方）的冷换流器被冷却并逐渐地沿竖井沉降，在其抵达底部并展开形成横贯室内的"冷空气池"之前对流效应不断加速。参见图 55。

重力供冷系统的优势在于完全没有噪声。配置该系统的房屋的制冷量随着冷却器处温差的不同而变化，并且可在一定程度上自动运行。只要保证顶部的空气入口与底部的空气出口设置合理，就能够在保证运行效果不受损的前提下，将重力供冷系统隐蔽的安装在墙衬或者窗帘后面。结露的问题可以参照前面所述系统中防结露的措施解决。关于主要冷传递系统的概述参见表 11。

**冷传递系统的特点** 表 11

| 冷传递系统 | 优点 | 缺点 | 输出能量比,辐射/对流 | $T_f[℃]$ |
|---|---|---|---|---|
| 冷却顶板 | 适宜的供水温度,快速可控性,辐射效应,房间可以独立控制 | 实际上无法排除凝水 | 50/50 | 10~16 |
| 循环空气供冷 | 对控制响应迅速,房间可以独立控制 | 纯对流效应,实际上无法排除凝水,风机噪声 | 0/100 | 6~10 |
| 重力供冷 | 对控制响应迅速,房间可以独立控制,完全无噪声,隐蔽安装 | 纯对流效应,凝水排水困难 | 0/100 | 6~10 |

### 3.4.3 混合系统

有很多传递系统适用于为室内供给热量或者冷量（参见表12）。

**空调设备**　这些系统也包括以集中形式为房间提供通风（或供热与供冷）的空调设备（参见图56）。空调系统的优势在于它能够对室内的空气进行加湿或者除湿，这意味着它几乎能够实现任何想要的室内空气条件。参见"通风系统"一章中的"机械通风"一节。

空气系统的一个基本缺点是空气并不是一个好的热传输介质，因为空气的蓄热能力很低（参见"设计原理"一章，"满足需求"一节）。增加空气流量来满足热量或冷量的需求，同时会导致输送气体使用的能量增加，而这一点在使用水系统时可以避免。空调系统中通常没有允许房间独立控制或者用户干预的设备。这一事实以及前面提到过的空气温度、流速和紊流强度对热舒适性的影响，均会导致用户的抱怨与不满意。

**热激活构件**　采用以辐射原理工作的系统通常能够提高舒适性（参见"设计原理"一章中的"舒适性需求"一节），例如采用热激活构件。在这些系统中，输送热媒或冷媒的盘管管线被浇筑在混凝土板的中心处（参见图57）。

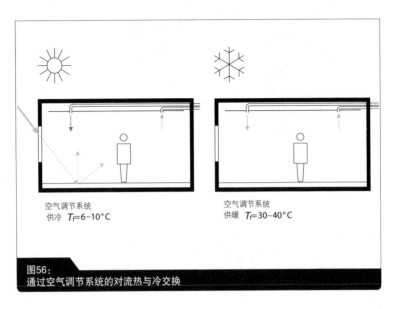

图56：
通过空气调节系统的对流热与冷交换

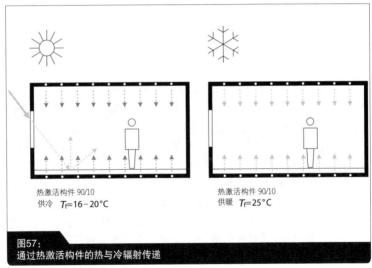

图57：
通过热激活构件的热与冷辐射传递

大块混凝土的热惰性导致可控性特别低。因此，热激活构件主要被用于满足恒定负荷或者满足基本水平的负荷。参见"设计原理"一章中的"满足需求"一节。

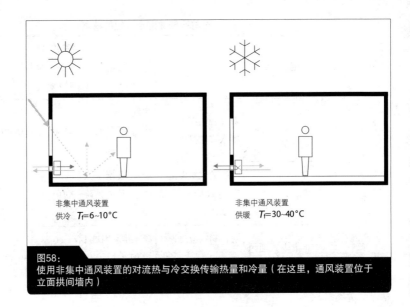

图58：
使用非集中通风装置的对流热与冷交换传输热量和冷量（在这里，通风装置位于立面拱间墙内）

用于供暖，供给温度为25℃。热激活构件的性能主要取决于顶棚表面与室内空气之间的温度差。这产生了一个自调节效应，即该系统冷量的输出随室内空气温度的升高而增大。

这类系统的露点问题与前述冷却顶板的相似，但在16~20℃这样高的冷媒供给温度工况下，不发生结露条件所允许的相对湿度会更高。供热与供冷两种模式下，适度的供给温度意味着这类系统可以采用可再生能源。

冷却介质与混凝土的热质量有效结合，可致使组件有效断电，例如在夜间。如夜间通风，这降低了（参见"设计原理"一章中的"满足需求"一节）白天所需的冷量输出，并且使全天的负荷分布图更加均匀，参见108页图14。

为使激活构件正常工作，混凝土表面必须是热易进入的，因此热激活顶棚系统与悬挂式顶棚是不相容的。此外，冲击声隔音也会降低这些系统的效力，因为其会使顶面的冷传输受到限制。

**非集中式通风装置**

非集中式通风装置被安置在立面（例如在拱间墙内）或者在立面区域（例如在双层龙骨地板内），并且与室外空气连通。参见图58。

这类装置的可控性非常好，但是空气调节设备需要增加流量来满足较高的冷负荷，这意味着仅用于供热或供冷时，非集中式通风

**热与冷传递混合系统的特点**　　　　　　　表 12

| 混合系统 | 优点 | 缺点 | 输出能量比,辐射/对流 | $T_f[℃]$ 供热 | $T_f[℃]$ 供冷 |
|---|---|---|---|---|---|
| 空气调节系统 | 对控制响应迅速；凝水容易排出 | 单纯对流效应；个性化房间控制选择差 | 0/100 | 30~40 | 6~10 |
| 热激活构件 | 适宜的供给温度；辐射效应；局部自动调节 | 容量限制；实际上无法排除凝水 | 90/10 | 25 | 16~20 |
| 非集中式通风装置 | 对控制响应迅速；容易实现个性化房间调节 | 凝水排除困难；增加运行维护成本 | 0/100 | 30~40 | 6~10 |

装置并不总是切实可行的选择。如果用分散式空调设备来满足较高的冷负荷，就需要采用一些排出凝水的方法。

分散式系统能够很好地与活性构件结合，因为两类系统间的优势可以互补，不足可以相互抵消。

当应用中央空调系统时，分散式机组供冷与供暖所需的供给温度分别为 6~10℃和 30~40℃。表 12 列出了最常见的热与冷传递混合系统的特点。更详细内容请参见附录第 164 与 165 页。

### 3.5 选择正确的系统

选择最合适的调温系统来满足所有需求，经常是设计师面临的一项艰巨的任务。存在多种能源，每种能源要求合适的热与冷的发生器，但并不是任一类发生器都能够与任一种传递系统相结合。各独立部件之间的相互作用复杂，以及或多或少可行的合理的组合数量多，致使与一个专业设计工程师的密切合作不可或缺。

| | 选择调温系统的标准 | 表 13 |
|---|---|---|
| 技术标准 | 容量（满足负荷需求）<br>与各构件匹配的系统温度<br>可再生能源的适用性<br>能源资源的可及性<br>热回收的可能性<br>可控性 | |
| 环境效应 | 主要能源需求<br>$CO_2$ 排放 | |
| 用户接受性（或认可度） | 舒适性要求<br>用户影响（用户干预） | |
| 经济性 | 初投资<br>运行费用 | |

系统的选择还与一系列进一步的标准相关，这些准则超出了纯粹的技术考虑。其中，最主要的是能源效率、用户认可度与经济可行性，所有这些意味着设计师必须非常仔细地进行设计。

表 13 概述了选择一个合适的调温系统所需考虑的最为重要的标准。

# 通风与调温的结合

## 4.1 可行方案的范围

所推荐的通风和调温系统必须能够切实的彼此结合,这样才能够开发出室内环境调节系统的整体概念,以确保能够达到需求的室内温度与通风。

低技术和高技术

根据需求,通风与调温有很多种可能的组合方法,这些方法的技术复杂程度不同。对室内环境调节,从技术上最简单(低技术)的由窗通风和散热器构成的变体起,直到最复杂(高技术)的全空气空调系统,均是其可能的概念范围。参见图59。

尽管这些系统代表两种极端类型,并因此而位于可能方案范围的始点和终点,然而,根据需求与使用的标准,它们仍可能是室内环境调节最为合适的概念,并为满足需求发挥作用。

从室内环境调节"专利方案"的意义上讲,系统部件可能组合的数量排除了任何普遍适用的组合。对于建筑师或者一个专业的设计工程师,为得到一个合适的室内环境调节的概念,更有成效的方法是以项目标准来评价各种可能的概念。

## 4.2 选择标准

通风、热与冷传递部件的组合可以根据技术标准(例如可再生能源以及热量回收)和用户认可度的相关标准(例如舒适程度以及用户影响)进行评估。

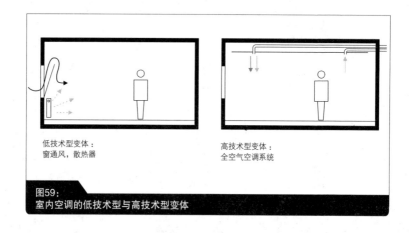

低技术型变体:
窗通风,散热器

高技术型变体:
全空气空调系统

图59:
室内空调的低技术型与高技术型变体

## 结语

　　建筑普遍需要个性化的室内环境调节解决方案。空调系统的每一个设备必须根据明确的适用工况范围与需求进行细致的检验，进而为建筑创造一个定制解决方案，这与其他行业形成了鲜明的对比。出于这个原因，实现一个成功的概念通常包含备选方案的比较测试。即便已经获得了一个满意的方案，也不应停止其他可能组合的检验以及它们优势和劣势的评价。

　　有经验的建筑师和工程师们能够更快地认识到有利和不利的方面，并借此推荐特殊的应用。然而，若没有更准确的分析，要掌握这一科目的难点（复杂性）是不可能的。

　　因此，本书力图以概述的方式来表明各独立系统组件的详细说明与一些组合的案例是非常重要的。这些内容提供了关于室内环境调节这一科目的介绍，以便今后能够通过计算以及技术图纸创建一个全面设计的室内环境调节概念。然而，只有通过理解其中的依存关系，并且认清具体的工程参数与技术可行性之间的相互作用，才能为项目找到一个最佳的解决方案。

# 附录

## 一些概念的案例

### 窗通风，散热器

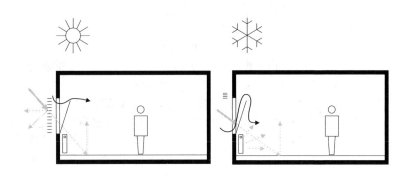

带散热器的窗通风 + 遮阳

| 适用性 | 优点 | 缺点 |
| --- | --- | --- |
| 居住建筑 | 通过水的高效能量输运 | 送风未经过调节 |
| 办公室 | 用户影响与个性化室内控制容易实现 | 不能确定换气次数 |
| | 实现通风与调温的解耦 | 不能进行热回收 |
| | | 不能冷却、除湿或加湿 |
| | | 冬季不舒适，热损耗大 |
| | | 可能引入室外释放的噪声与灰尘 |

窗通风，地板供暖

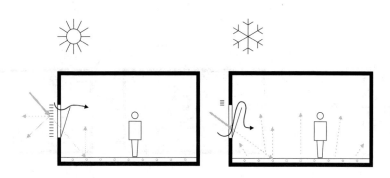

带地板供暖的窗通风 + 遮阳

| 适用性 | 优点 | 缺点 |
| --- | --- | --- |
| 居住建筑 办公室 | 通过水的高效能量输运 适宜采用可再生能源 用户影响与个性化室内控制容易实现 舒适的热传递 实现通风与调温的解耦 | 送风未经过调节 不能确定换气次数 不能进行热回收 不能冷却、除湿或加湿 冬季不舒适，热损耗大 可能引入室外释放的噪声与灰尘 对控制响应很迟钝 |

空气供暖

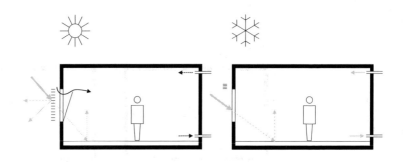

带夏季选择性通风的空气供暖 + 遮阳

| 适用性 | 优点 | 缺点 |
| --- | --- | --- |
| 居住建筑 | 无附加供暖面积需求 | 通过空气输运能量的效率低 |
| | 可确定换气次数 | 仅在低热量需求时有效 |
| | 能进行热回收 | 单一对流传热 |
| | 舒适的热传递 | 很难融入用户影响与个性化室内控制 |
| | 可避免室外释放的噪声与灰尘 | 通风与调温耦合 |

窗通风，对流式散热器，冷辐射吊顶

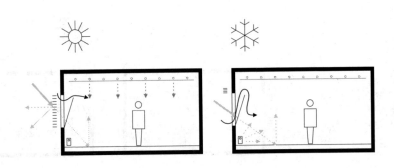

窗通风、对流式散热器、冷辐射吊顶 + 遮阳

| 适用性 | 优点 | 缺点 |
| --- | --- | --- |
| 办公室 | 夏季舒适性高 | 送风未经过调节 |
| 会议室 | 通过水的高效能量输运 | 不能确定换气次数 |
| | 用户影响与个性化室内控制容易实现 | 不能进行热回收 |
| | 实现通风与调温的解耦 | 不能冷却、除湿或加湿 |
| | | 冬季不舒适且热损耗大 |
| | | 可能引入室外释放的噪声与灰尘 |

窗通风，对流式散热器，循环空气机组

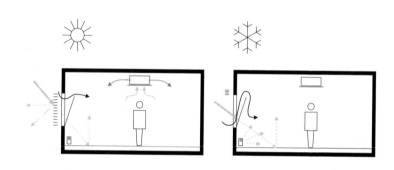

窗通风、对流式散热器、循环空气机组 + 遮阳

| 适用性 | 优点 | 缺点 |
|---|---|---|
| 办公室 | 用户影响与个性化室内控制容易实现<br>实现通风与调温的解耦 | 送风未经过调节<br>不能进行热回收<br>仅对流方式的热与冷的传递<br>冬季可能不舒适且热损耗大<br>可能引入室外释放的噪声与灰尘 |

非集中通风系统，热激活组件

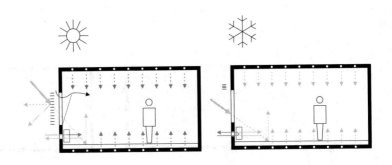

带夏季选择性窗通风的非集中通风系统、热激活组件 + 遮阳

| 适用性 | 优点 | 缺点 |
|---|---|---|
| 办公室 | 通过水的高效能量输运（满足基本负荷） | 安装费用很高 |
| 会议室 | 部分热与冷通过辐射传递 | 凝水排除困难 |
|  | 送风能够进行预处理 | 不能加湿与除湿 |
|  | 可以确定换气次数 | 维护成本高 |
|  | 能够进行热回收 |  |
|  | 可能实现用户影响与个性化室内控制 |  |
|  | 实现通风与调温的解耦 |  |
|  | 可避免室外释放的噪声与灰尘 |  |

空气调节系统

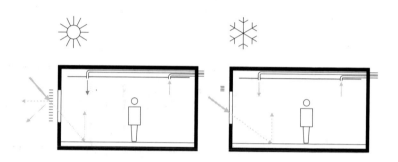

空气调节系统 + 遮阳

| 适用性 | 优点 | 缺点 |
|---|---|---|
| 住宅 | 送风能够进行预处理 | 通过空气输运能量的效率低 |
| 办公室 | 能够确定换气次数 | 单一对流方式的冷与热传递 |
| 会议室 | 能够供暖、供冷、加湿和除湿 | 用户影响与个性化室内控制难以实现 |
|  | 能够进行热回收 | 通风与调温耦合 |
|  | 容易排除凝水 |  |
|  | 可避免室外释放的噪声与灰尘 |  |

　　下表概述了本书中介绍的各种通风系统与调温系统的可能组合。这个表既不普遍适用也不详尽，但应该将其作为一个展示如何比较各种概念以及找到可行的解决方案的例子。所选的作为例子的各个概念也在本书的附录中予以详细的展示。

| 室内环境调节系统 | 通风方式 | | 调温方式 | | | | | | 特点 | | | | 典型应用 | | |
|---|---|---|---|---|---|---|---|---|---|---|---|---|---|---|---|
| | 自然 | 机械 | 散热体 | 表面供暖 | 空气供暖 | 冷辐射吊顶 | 再循环空气供冷 | 空气调节系统 | 热激活组件 | 非集中通风机组 | 适用于可再生能源 | 热回收 | 可控性 | 舒适性 | 用户影响 | 住宅 | 办公室 | 会议室 |
| (图) | X | | X | | | | | | | | ○ | − | + | ○ | + | X | X | |
| (图) | X | | | X | | | | | | | + | − | − | + | + | X | (X) | |
| (图) | (X) | X | | | X | | | | | | − | + | + | ○ | − | X | | |
| (图) | X | | X | | | X | | | | | ○ | − | + | + | + | | X | X |
| (图) | X | | X | | | | X | | | | ○ | − | + | ○ | + | | X | X |
| (图) | (X) | X | | | | | X | X | | | ○ | + | ○ | + | + | | X | X |
| (图) | | | X | | | | X | | | | − | + | + | − | − | (X) | X | X |

## 标准

### 本书参考的标准与指南

| | |
|---|---|
| DIN 1946 | "Ventilation and air conditioning", Part 2 "Technical health requirements (VDI ventilation rules)", 1994-01 (withdrawn) |
| DIN EN 12831 | "Heating systems in buildings - Method for calculation of the design heat load", Supp.1 "National Annex", 2006-09 |
| DIN EN 15251 | "Indoor environmental input parameters for design and assessment of energy performance of buildings addressing indoor air quality, thermal environment, lighting and acoustics", 2007-08 |
| DIN EN ISO 7730 | "Ergonomics of the thermal environment - Analytical determination and interpretation of thermal comfort using calculation of the PMV and PPD indices and local thermal comfort criteria (ISO 7730:2005)", 2006-05 with Amendment 2007-06 |
| DIN V 18599 | "Energy efficiency of buildings - Calculation of the net, final and primary energy demand for heating, cooling, ventilation, domestic hot water and lighting", Part 1 "General balancing procedures, terms and definitions, zoning and evaluation of energy sources", 2007-02 |
| VDI 2078 | Technical regulation "Cooling load calculation of air-conditioned rooms (VDI cooling load regulations)", 1996-07 |
| VDI 3803 | Technical regulation "Air-conditioning systems - Structural and technical principles", 2002-10 |

## 参考文献

Hazim B. Awbi: *Ventilation of Buildings*, E & FN Spon, London 1991
Sophia and Stefan Behling: *Solar Power*, Prestel, Munich 1996
Klaus Daniels: *Advanced Building Systems*, Birkhäuser Verlag, Basel 2003
Klaus Daniels: *Low-Tech Light-Tech High-Tech*, Birkhäuser Verlag, Basel 1998
Baruch Givoni: *Climate Considerations in Building and Urban Design*, John Wiley & Sons, New York 1998
Baruch Givoni: *Passive and Low Energy Cooling of Buildings*, John Wiley & Sons, New York 1994
Gerhard Hausladen, Michael de Saldanha, Petra Liedl, Christina Sager: *Climate Design*, Birkhäuser Verlag, Basel 2005
Norbert Lechner: *Heating, Cooling, Lighting: Design Methods for Architects*, John Wiley & Sons, New York 1991
Christian Schittich (ed.): *Solar Architecture*, Birkhäuser Verlag, Basel 2003
Steven V. Szokolay: *Environmental Science Handbook for Architects and Builders*, The construction press/Lancaster, London 1980
Steven V. Szokolay: *Introduction to Architectural Science: The Basis of Sustainable Design*, Elsevier Architectural Press, Oxford 2004

## 作者简介

奥利弗·克莱因（Oliver Klein），理学学士，是一位建筑师与能源顾问。2005年起，在德国多特蒙德理工大学（Dortmund University of Technology）的气候可兼容建筑学院（Faculty of Climate-compatible Architecture）任研究生助理，主要从事教学与科研工作。

约尔格·施伦格尔（Jörg Schlenger），理学学士，是一位结构工程师及能源与空调概念顾问，擅长建筑热工过程模拟。2004年起，在德国多特蒙德理工大学的气候可兼容建筑学院任研究生助理，主要从事教学与科研工作。

## 译者简介

马志先，工学博士，大连理工大学土木学院建筑能源研究所，从事暖通空调系统传热传质与强化传热试验和理论基础研究，以及建筑环境测试技术教学。